**ABOU TANGARA**

# SPATIO-TEMPORAL DYNAMICS OF INVASIVE AQUATIC PLANTS

ABOU TANGARA

# SPATIO-TEMPORAL DYNAMICS OF INVASIVE AQUATIC PLANTS

## The contribution of remote sensing and GIS to mapping the spatio-temporal dynamics of aquatic plants

ScienciaScripts

**Imprint**
Any brand names and product names mentioned in this book are subject to trademark, brand or patent protection and are trademarks or registered trademarks of their respective holders. The use of brand names, product names, common names, trade names, product descriptions etc. even without a particular marking in this work is in no way to be construed to mean that such names may be regarded as unrestricted in respect of trademark and brand protection legislation and could thus be used by anyone.

Cover image: www.ingimage.com

This book is a translation from the original published under ISBN 978-620-6-72087-4.

Publisher:
Sciencia Scripts
is a trademark of
Dodo Books Indian Ocean Ltd. and OmniScriptum S.R.L publishing group

120 High Road, East Finchley, London, N2 9ED, United Kingdom
Str. Armeneasca 28/1, office 1, Chisinau MD-2012, Republic of Moldova, Europe
Printed at: see last page
**ISBN: 978-620-8-06676-5**

## DEDICATION

***This book is dedicated to :***

*My father for instilling in me a love of work;*

*My mother for her blessings;*

*My Uncle for the encouragement;*

*My Auntie for her advice;*

*My brothers and sisters ;*

*All the TANGARA, TRAORÉ and DEMBÉLÉ families for their moral support.*

***May the Almighty bless you***

**ACKNOWLEDGEMENTS**

We would first like to pay tribute to all those who have contributed to its success.

We would like to express our deepest respect and gratitude to :

- Dean of the UFR of Earth Sciences and Mining Resources (STRM), prof. SORO Nangnin for accepting us into the UFR STRM;
- CURAT Director, Dr KOUAME KAN Jean, for the quality of the training provided at CURAT;
- Doctor YOUAN TA Marc, Scientific Director of this dissertation, for his confidence in accepting the direction of this work;
- Doctor DIBI N'Da Hyppolite, our co-director, to whom we say a big thank you. Despite your busy schedule, you were present for the framing of the work. Your enthusiasm, patience and advice have left their mark on this work. It was a real pleasure for us to work under your supervision;

We would also like to thank all the CURAT staff for their availability;

To all my classmates of the 2017-2018 promotion for their collaborations;

We warmly thank the members of the jury: Dr. N'Guessan Yao Alexis, Dr. Komoe Koffi for the honour of judging this work.

We would also like to thank our parents, siblings and friends. Many thanks to all of you.

We cannot conclude without expressing our gratitude to all the teachers at the History and Geography Faculty of the University of Social Sciences and Management of Bamako in Mali. In particular, we would like to thank Dr. Hamadoun TRAORÉ, head of the Geography department, and Dr. Baba Faradji N'DIAYE, teacher-researcher, for making recommendations for this study. Thank you very much.

Finally, for all the people mentioned or not, we say this prayer to you: "May GOD the all-powerful, all-knowing and all-present, who has witnessed all that you have done for me, grant you a greater reward according to his will. Amen"

CONTENTS

## SUMMARY

In Mali, the river is 1750 km long. It flows through six (6) of the country's major cities: Bamako, Koulikoro, Ségou, Mopti, Timbuktu and Gao. The river is of ecological, cultural, social and economic value to the local population. It also plays a key role in maintaining biodiversity and in farming, fishing and tourism. Despite its many functions, the river is subject to heavy colonisation by invasive aquatic vegetation (IAV), which hinders its proper functioning.

Remote sensing is the tool we used to monitor the dynamics of the proliferation of invasive plants on the riverbed in the District of Bamako. Landsat ETM+ and OLI satellite images were used to carry out this study by monitoring the spatio-temporal dynamics of EABs. In total, we used (12) images, i.e. (3) images per year, all downloaded from "earthexplorer.usgs.gov" from 2000 to 2018.

In the methodological approach, after acquiring the satellite images, we carried out pre-processing, image enhancement or processing and then cartographic rendering. The intra- and inter-annual dynamics were also studied, followed by an assessment of the impact of the VAEs on the flow of the river from 2000 to 2018.

Using a multi-date diachronic analysis, the study shows that the dynamics of the proliferation of EABs on the river vary from one year to the next. Over the period 2000 - 2018, the study shows a decrease in EABs from March to September, but also inter-annual growth, with a spatial expansion rate of 15.50%. We also studied the relationship between changes in EAVs and river flow, and the curve representing EAVs showed a decrease in EAVs as river flow increased. Finally, after assessing the impact of the VAEs on the flow of the river, it was found that the VAEs did not reach the average that could prevent the river from flowing. In view of some of the study's limitations and with a view to proposing ways of managing VAE, a series of recommendations and prospects were formulated. Some of the recommendations concern raising public awareness of the effect of river water pollution on the proliferation of WAS. In the future, it will be interesting to produce a map of the dynamics of land use as a likely source of nutrients for AEVs.



**Key words:** Remote sensing, GIS, VAE, spatio-temporal dynamics, Niger River, Landsat.

## INTRODUCTION

The River Niger is the third largest river in Africa (after the Nile and the Congo), both in terms of length (4,200 km) and surface area (2,000,000 km)[2)] . According to Cissé *et al* (2007), the river lies at the heart of West Africa and part of Central Africa. It rises in Guinea and flows through Mali, Niger, Benin and Nigeria before emptying into the Atlantic Ocean.

In recent years, the River Niger has come under considerable threat from climatic hazards, demographic pressure and anthropogenic pollution. This will compromise the sustainability of the resource in terms of both quality and quantity (ABN, 2002).

The current situation of the river and its tributaries is of increasing concern to many stakeholders involved in the sustainable development of its resources: users, developers, water and environmental managers and political decision-makers. The quality of the water in the River Niger is becoming increasingly unhealthy. (ABFN, 2004).

In the Malian section, the River Niger runs for 1,750 km and is home to almost 80% of the population, who live mainly from farming, livestock rearing and fishing (Cissé *et al,* 2007). The river flows through no less than 6 of Mali's major cities, namely : Bamako, Koulikoro, Ségou, Mopti, Timbuktu and Gao. Its contribution to the Malian economy is 101,008 tonnes of fish produced annually and 671,657.77 tonnes of agricultural production in the irrigated areas of the River Niger alone (Sidibé S., 2017). This river is considered as a nourishing artery for the country that maintains living conditions for the populations. (https://maliactu.net/mali-fleuve-niger-au-mali-enjeux-et-perspectives/ 29/10/2018).

In Mali, an inventory of floating plants in the bed of the River Niger has shown that the water hyacinth poses serious problems in the Koulikoro region and the District of Bamako (Cheickna *et al,* 2004). The problem of invasive aquatic plants is becoming increasingly worrying in Mali in general and in Bamako in particular, where their management requires considerable material, human and financial resources.

It is clear that invasive aquatic plants are a considerable drain on the economies of the countries where they are present, due to the rate at which they spread. Faced with this scourge, various initiatives have been taken by the Malian government's technical services and by the local population. Most of these initiatives have been based mainly on manual removal and biological control.

In 2002 and 2003, VAE survey work was carried out by IER (Institut d'Économie Rurale), as part of the project to combat water hyacinth in the bed of the River Niger. A study of invasive aquatic plants was also carried out by GERACT-Sarl in 2004 and 2005, funded by the ABFN (Agence du Bassin du Fleuve Niger). As part of the implementation of its special investment budget for 2004, the Agency initiated a restricted consultation in order to better control VAE in the River Niger. These interventions provided an initial idea of the sites of permanent infestation of VAE in Mali.

To work in this direction, while focusing on the management of EAB, we have chosen remote sensing to monitor the spatio-temporal dynamics of these invasive aquatic plants in the bed of the River Niger in the District of Bamako. The aim was to obtain an up-to-date map of EABs so that we could better monitor their dynamics and devise strategies to reduce their proliferation in the bed of the River Niger. To better monitor this spatial proliferation, it is important to use earth observation data. Controlling the spatial dynamics of invasive aquatic plants is an important basis, hence the interest of our study topic, which is entitled: "Contribution of remote sensing and GIS to mapping the spatio-temporal dynamics of invasive aquatic plants in the bed of the River Niger: the case of the District of Bamako (Mali)".

The general aim of this study is to contribute to a better understanding of invasive aquatic plants in the bed of the River Niger in the District of Bamako.

Specifically, this will involve :

- Produce a spatio-temporal map of the spatial distribution of invasive aquatic plants in the bed of the River Niger near Bamako between 2000 and 2018;
- Determine the rate of spatial expansion of invasive aquatic plants in the riverbed between 2000 and 2018;
- analyse the impact of the proliferation of invasive aquatic plants on the flow of the river over the period 2000 - 2018.

This dissertation is in three (3) parts: The first part presents general information on the study area, invasive aquatic plants and the concept of remote sensing-GIS; the second, the methodological approach and materials used during the work and finally, the last part summarises the results and discussion.

## CHAPTER 1: PRESENTATION OF THE STUDY AREA

### 1.1 Physical study

#### 1.1.1 Geographical context

Bamako, the economic capital of Mali, is situated on the banks of the River Niger between 12°34'30" and 12°40'00" north latitude and 8°20'00" and 07°56'30" west longitude. Bamako is the country's main crossroads, thanks to its position as the point of convergence of the main roads linking the various regions of Mali. In 2009, the city had a population of 1,809,106 (INSTAT, 2009). The city is built in a basin surrounded by hills. It stretches from west to east over 22km² and from north to south over 12km², covering an area of 267km$^2$ . The district of Bamako is subdivided into six (6) communes: (Commune I, II, III, IV, V and VI). Figure 1 shows the study area (District of Bamako).

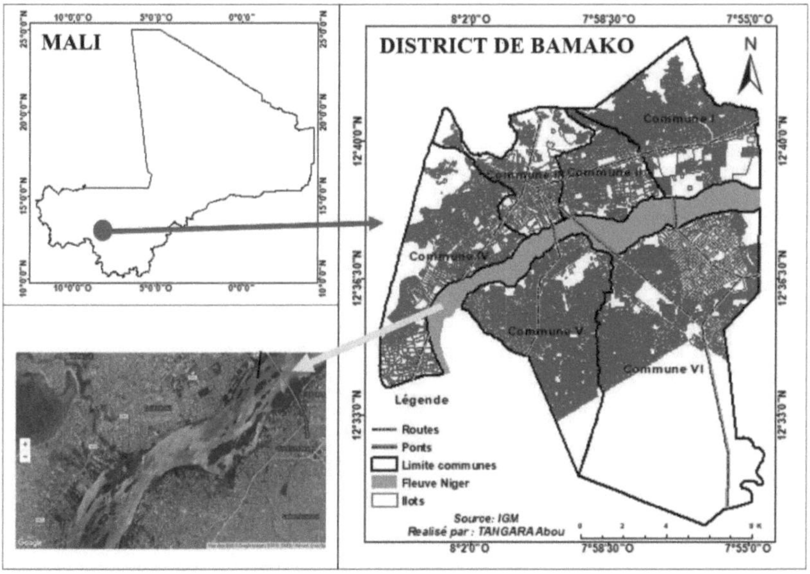


**Figure 1 Map of the study area .**

#### 1.1.2 Vegetation and climate

The vegetation cover of Bamako depends on spatial planning policies. The vegetation is tropical, with gallery forests along watercourses (Millimono, 2009). There are a number of tree species, including néré (*Parkia biglobosa*), caicedra (*Khaya senegalensis*), acacia (*Acacia senegalensis*), mango (*Mangifera indica*) and nyme (*Azadirachta indica*).

The Bamako area enjoys a humid tropical climate with high temperatures (35.10°C to 22°C). Figure 2 below graphically represents the average monthly rainfall and temperature values in the city of Bamako. This climate is characterised by the alternation of two seasons, a wet season lasting from 6 to 9 months and a short dry season. The monthly variation in the number of rainy days is between 0 and 24 days, with the maximum occurring in August.

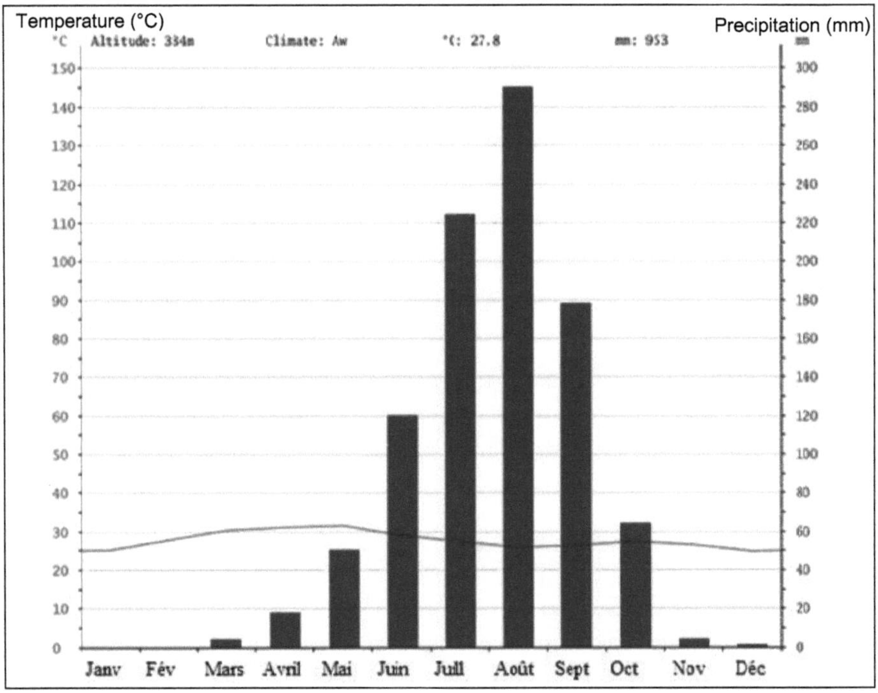


**Figure 2 Umbro-thermal diagram for Bamako (Service Météo 2010)**

Source : https://images.climate-data.org/location/500/temperature-graph.png

### 1.1.3 Relief and soil

With a relatively flat topography, Mali is made up of plateaux and plains, with well-developed dune systems in the north and east (Maïga, 2012).

The relief of Bamako is characterised by a predominance of sandstone plateaux and granite hills. Its morphology is linked to its geology. As this part of Africa has only undergone insignificant tectonic movements, the relief is mainly made up of geological layers that have been exposed by differential erosion (Diarra, 2003). Figure 3 shows the relief of Bamako.

The district of Bamako is located on the plains of the Niger. As a result, most of the soils are made up of fine clay, sandy loam and hard sandstone in some places (Maïga, 2012).

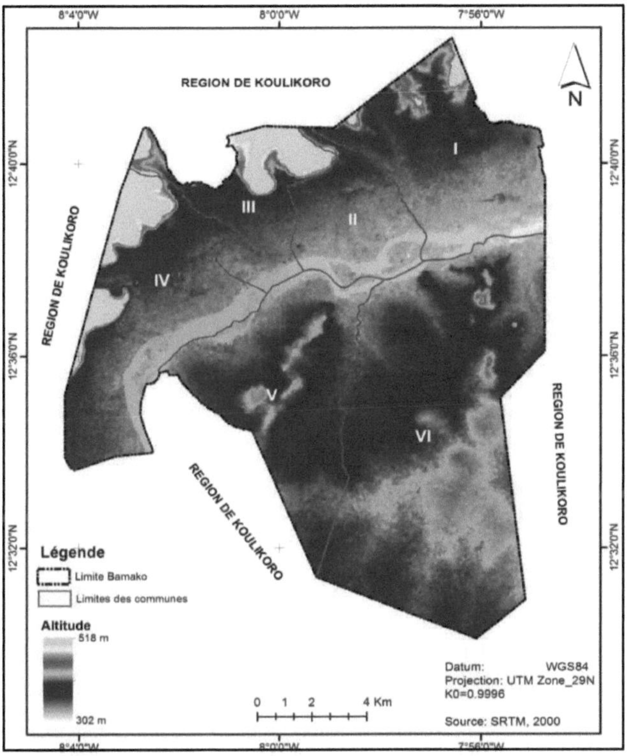


**Figure 3 Relief map (Source: SRTM, 2000)**

### 1.1.4 Hydr ography

The hydrology of the District of Bamako is marked by surface water, dominated by the River Niger. The river divides the city of Bamako into left and right banks. The river's flow varies logically with regional rainfall. It peaks during the winter season in August-September and falls to its lowest point during the dry season. Figure 4 shows the river's average monthly flow. Data produced by the Bamako hydrological station. The river is of vital importance to the local population, as it is used for market gardening, fishing, irrigation, sand mining and dyeing.

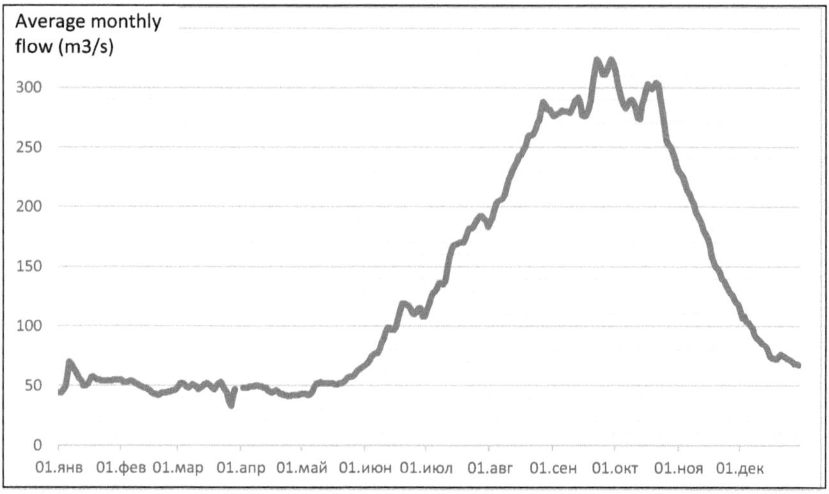


**Figure 4 Average monthly flows (in m3/s) in 2017. Bamako hydrological station Source:** Direction National d'Hydraulique (2017)

### 1.2 Human environment

#### 1.2.1 Historical overview

The origins of Bamako date back to 1640, when the city was named after BAMBA SANOGO, a hunter who occupied the site with his hunting camp (Maïga, 2012).

The development of Bamako can be explained by the geographical position of this small village of four districts (Bagadadji, Niaréla, Bozola and Dravéla), situated at the crossroads of the North and South, and therefore of different civilisations (Diarra, 1999).

#### 1.2.2 Demographic context

The District of Bamako covers a total area of around 18,000 hectares. The city experienced strong growth of 5.97% between 1997 and 2002 (INSTAT, 2009).

Some of the fairly strong population growth in Bamako can be explained by the relatively high fertility rate of Bamako women and also by the fall in the mortality rate.

However, migration plays a dominant role. Population growth in the outskirts of Bamako is fuelled by migration. 87% of migrants come from within the country. The average age of migrants when they arrive in Bamako is around 20 (Maïga, 2012). Urbanisation in Mali shows major regional disparities. In 2004, Bamako accounted for 32.7% of the country's total urban population. The annual population growth rate in Bamako was estimated at around 5.8% (Konaté, 2011). Bamako's demographic dynamic has been accompanied by a spatial dynamic. The emergence of spontaneous neighbourhoods is considered to be a revealing sign of urbanisation (Konaté, 2011). Figure 5 shows the demographic evolution of Bamako by commune.

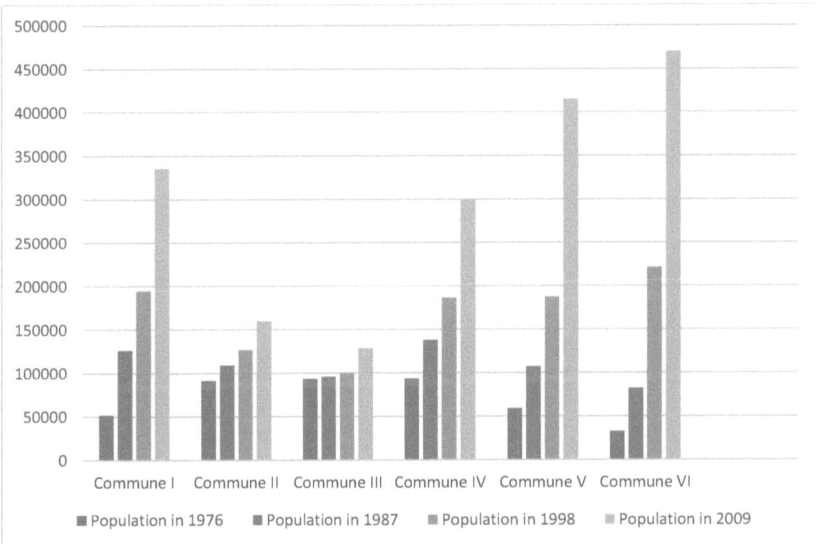


**Figure 5 Demographic trends in Bamako's communes from 1976 to 2009**

**Source:** National Statistics and Information Directorate (2009)

### 1.2.3 Economic framework

The economy of the District of Bamako is characterised by the agricultural and commercial sectors.

The agricultural sector is dominated by market gardening. Income from market gardening plays a decisive role in the economic equilibrium of the family farm.

According to the results of the 1988-1989 budget-consumption survey (DNSI, 1994), Bamako is now a major consumer of market garden produce, particularly fruit and vegetables.

Trade in the District of Bamako is dominated by small retailers based in shopping centres and neighbourhoods. Wholesalers and semi-wholesalers are concentrated mainly in the city centre. The largest are those selling motorbike spare parts, food, hydrocarbons and fabrics. On the basis of patented production, the District of Bamako accounts for around 70% of commercial activity and over 86% of the country's imports and exports (Maïga, 2012).

# CHAPTER 2: GENERAL INFORMATION ON INVASIVE AQUATIC PLANTS (IAP) AND THE CONCEPT OF REMOTE SENSING-SIG

## 2.1 Characteristics of invasive aquatic plants (VAE)

### 2.1.1 Water hyacinth (*Eichhornia crassipes* )

The water hyacinth known as *Eichhornia crassipes* is a floating and invasive aquatic plant. According to Qaiser *et al* (2005), it is a perennial herbaceous plant belonging to the spermaphytes phylum, the angiosperms subphylum (*Magnoliophyta*), the monocotyledons class (*Liliopsida*), the *Commelinidae* subclass, the *Pontederiales* order, *Pontederiaceae* family.

#### 2.1.1.1 Origin of the water hyacinth

According to Barret and Forno (1982), the hyacinth originated in South America. It was from this area that it was accidentally or intentionally introduced by man to various parts of the world. It was subsequently found in North and Central America, Asia, India, Australia and New Zealand.

In Africa, it was first reported in the Nile Delta at the beginning of the 20th century (Holm *et al,* 1977). In Mali, its presence dates back to 1990 around the town of Bamako (Dembélé, 1994). From this point onwards, it has invaded the course of the River Niger and represents a threat to the country's water bodies.

#### 2.1.1.2 Description of the water hyacinth

Water hyacinth is a free-living, floating, stoloniferous hydrophytic plant. Morphologically, the water hyacinth can vary considerably in shape, colour, leaves and flowers (Dagno *et al,* 2007). It can reach 30 cm to 1 m in height (Fortier, 2007). The leaves are leathery with a cordate blade with parallel veins of a shiny light green colour (IER, 2012).

The petiole (5 cm in diameter and 30 to 50 cm long) is a tapered float under conditions of high plant density. Seeds are produced in large numbers in capsules, and each capsule can contain up to 300 seeds (Rutabagaya, 2017).

The inflorescence is a terminal spike, consisting of 10 to 30 flowers with 6 violet-blue or violet-pink petals, with a yellow mark in the centre. The roots are black, pendulous and long, forming below the rosette formed by the leaves at the base of the nodes.

The roots have many absorbent hairs. According to Ranarijaona *et al* (2013) and Pocquet *et al* (2015), they are resistant to long periods of drought and germinate as soon as they are immersed in water.

#### 2.1.1.3 Plant ecology and biology

Water hyacinth is a fast-growing floating plant par excellence. It also has the ability to adapt to life on land in the event of severe low water. Hyacinths multiply mainly vegetatively from stolons dispersed by currents (Navarro and Phiri, 2000). Vegetative propagation is very important in new infestation sites. New plants are produced from the elongation of stolons caused by the division of axillary meristems on the mother plant (Dagno *et al,* 2007). The very fragile clones remain attached to the mother plant by the stolon, then detach under the pressure of water currents, allowing new individuals to colonise other areas (Oumarou *et al,* 2008). The hyacinth produces seeds in large quantities, 5,000 seeds per plant (Navarro and Phiri, 2000). According to Gopal (1987), seeds are considered to be the plant's main multiplication factor. The flower-bearing stem tilts in the water and the seeds sink to the bottom where they can live for several years (15 to 20 years) without losing their germinative power (Rutabagaya, 2017).

### 2.1.2 Water fern or Salvinia molesta

The second invasive aquatic plant that we spotted on the river in the District of Bamako is the water fern. It is less common than the water hyacinth, but is a danger to the river.

#### 2.1.2.1 Origin and history

This plant is native to south-east Brazil. Between 1972 and 1990, it gradually invaded many rivers in America, Asia and Australia. Its presence has been reported in more than 20 countries (Sané, 2001).

In Mali, *Salvinia molesta* was observed in 2000 on the right bank of the River Niger, in the Bamako district (Dembélé, 1994).

#### 2.1.2.2 Description of *Salvinia molesta*

It is characterised by rounded green leaves that float on invaded bodies of water. The upper surface of the leaves is invaded by a network of long, rigid, impermeable hairs that allow the leaves to float. The umbrella shape of the top of these hairs is a specific feature (ECOWAS, 1996).

#### 2.1.2.3 Plant biology and ecology

*Salvinia molesta* spreads very quickly by vegetative reproduction from auxiliary buds. It is an aggressive and competitive plant. The plant can completely cover bodies of water with a thick carpet. As plant density increases, it spreads into pits and basins in slow-flowing rivers and irrigation canals (Dembélé *et al,* 1999).

## 2.2 Proliferation of invasive aquatic plants

Water pollution is the main cause of the proliferation of invasive aquatic plants. The degree of infestation of a body of water by harmful floating plants depends essentially on the availability or lack of nutritional requirements and the ecological conditions favourable to their proliferation (Ouédraogo, 2000).

In the city of Bamako, hyacinths proliferate where agricultural, industrial and household pollutants are discharged into the river via rainwater or wastewater drainage channels, without any form of treatment (Figure 6 below). This enriches the river with nitrogen and phosphorus, the main components of organic matter. This environment is ideally suited to the proliferation of the plant.

Another very favourable factor for the proliferation of invasive aquatic plants is the absence of natural enemies and also the absence of competitive plants. The water hyacinth's high reproductive capacity forms a thick floating carpet that prevents light from reaching submerged plants, thereby preventing photosynthesis (IER, 2012). It easily replaces native aquatic plants.

The danger of the proliferation of invasive aquatic plants is also represented by the difficult access to water for local populations.

**Figure 6Rubbish dump on the river bank in the Bozola district (Bamako)**. Source: (Photo TANGARA Abou, September 2018)

## 2.3 Different ways of combating invasive aquatic plants

**- Biological control**

Biological control is the use of living organisms and their products to prevent or reduce losses or damage caused by pests (Jiménez and Maricella, 2016). Biological control of weeds is based on the use of natural enemies of the host, so as to reduce the population to limits where it does not cause economic damage. Several organisms have been studied for the biological control of water hyacinth infestations. In Mali, two weevil species have been used for biological control. These are the beetles *Neohydronomus affinis* and *Neochetina bruchi* (IER, 2012).

**- Physical (mechanical) control**

Mechanical removal of aquatic plants is generally considered to be the best short-term solution. It is expensive, however, as it involves the use of land-based clamshell cranes. Draglines or water-based machines such as mowers and floating dredgers can also be used. Despite this, these methods are only suitable for relatively small areas. These techniques need to be supported by vehicles travelling on land or water to transport the large quantities of weeds removed (ECOWAS, 2003). The relatively high cost of purchasing and maintaining these machines could compromise the positive but limited and temporary effects of the method.

**- Physical control (manual)**

This is the main control method used in Mali (IER, 2012). Manual control is possible in easily accessible, shallow areas such as irrigation canals and ditches. The best time for manual control is the dry season (January to March). It must be carried out in collaboration with the local population. Only their participation can guarantee the sustainability of this form of control.

**- Chemical control**

This technique involves applying chemical substances (herbicides) that alter the metabolism and growth of plants, causing them to die.

This form of control is a last resort. The disadvantages of this form of control are mainly environmental and health-related, particularly where people drink the infested water. The method does not offer a permanent solution to weed infestation.

**- Biological and manual integrated control (recommended)**

It will focus on biological control to find a permanent solution, while physical control will be used as a short-term solution to small, localised infested areas. Physical pest control will be used to reinforce biological control on a regular basis, thereby keeping the problem under control as much as possible in the shorter term. Integrated pest management will be supported by the treatment of waste effluents and the monitoring of water quality (management of water catchments), to reduce the nutrients added to the water (ECOWAS, 1996).

The public awareness campaign on the possible use of invasive aquatic plants after their removal could encourage the participation of the local community in the control programme. Mechanical control will be considered if the costs and maintenance of the machines are bearable by the country. If this is not the case, manual control is recommended in the integrated pest management programme, when the mechanical method risks isolating the participation of local communities, particularly women, in weed control (ECOWAS, 1996).

## 2.4 Negative and positive impacts of invasive aquatic plants

### 2.4.1 Negative impacts s

One of the most remarkable consequences for the river is eutrophication. This is a natural process whereby bodies of water receive large quantities of nutrients, particularly phosphorus and nitrogen (Hade, 2002). Excessive growth of aquatic plants forms a green layer on the river, suggesting that underneath lies a body of water with aquatic life endangered by eutrophication.

***Environmentally:*** it causes a substantial change in the physical and chemical properties of water (Castillon, 2005). It pollutes the atmosphere because its anaerobic decomposition in water emits methane, a powerful greenhouse gas (GHG) (Traoré, 2006).

***Economically:*** it poses a serious threat to agricultural production, by blocking irrigation channels and drainage systems. It harms fishing by clogging fishing nets and sharply reducing oxygen levels in the water and therefore fish yields. It also reduces the activity of hydroelectric power stations, compromising the amount of electricity available (ECOWAS, 2003).

***Health:*** it creates a favourable environment for mosquitoes and encourages the development of certain diseases, such as malaria and bilharzia (Navarro and Phiri, 2000).

### 2.4.2 Positive impacts

Although it is harmful to the ecosystem, water hyacinth also has properties that can be exploited. According to studies by Saunders (2013) carried out in China, water hyacinth is used to eliminate unpleasant odours. Castillon (2005) has also shown that the compost produced from this aquatic plant is a fertiliser that gives a better yield and is less expensive than chemical fertilisers. It is also a feed for livestock (Traoré, 2006).

## 2.5 Physico-chemical parameters influencing the development of invasive aquatic plants

- **Physical parameters**

The physical parameters considered are temperature, conductivity, pH, dissolved oxygen and the depth of the water body (Justine, 2017). Hyacinth growth is very strong during periods of high water, when the plant floats on the water (Komelan, 1999) and flourishes at a temperature of 20 to 30°C. Its pH tolerance is estimated at 5 to 7.5 (Jardin, 2015).

- **Mineral parameters**

The minerals that play a role in the development of invasive aquatic plants, particularly hyacinth, are nitrates, phosphates, potassium and sodium (Justine, 2017).

- **Organic parameters**

The availability of organic matter in water, combined with other factors, encourages the growth of invasive aquatic plants. Excess organic matter is a source of pollution and asphyxiation in the aquatic environment (Hassane, 2010).

## 2.6 Earth observation data and the study of invasive aquatic plants

Remote sensing is considered to be a source of spatial data. Satellite images make it possible to systematically measure and monitor the state of plant cover at different spatial and temporal scales (Jacquin, 2010).

Invasive aquatic plants were mapped by Jean (2018), in the study of variation in the dynamics of invasive aquatic plants on Lake Tengréla in Burkina Faso, using remote sensing. This study was carried out using a series of MODIS satellite images from 2000 to 2017. The results of this study were satisfactory, as the MODIS satellite was able to highlight the occupancy stations of invasive aquatic plants on the lake.

Remote sensing offers many possibilities for mapping, analysing and managing invasive aquatic plants. Given the importance of remote sensing in spatial analysis, this study can be useful at several levels:

- to understand the colonisation dynamics of invasive aquatic plants;
- to plan intervention measures to combat aquatic plants during their regression phase;
- decision-making support for invasive aquatic plant management strategies.

For this, the role of remote sensing in the dynamics of invasive aquatic plants is undoubtedly essential. Spatial and temporal analysis using earth observation data shows the various changes that have taken place in the fight against invasive aquatic plants.

## 2.7 THE CONCEPT OF REMOTE SENSING AND GIS

### 2.7.1 Definitions of remote sensing :

Remote sensing from space is a scientific discipline that integrates a wide range of skills and technologies used for the observation, analysis and interpretation of terrestrial and atmospheric phenomena based on measurements and images obtained using aerial and space platforms. It involves the acquisition of information at a distance, without direct contact with the object being studied (Francisco *et al,* 2013).

According to Boulerie (2008). "It is the body of knowledge and techniques required to interpret various "objects" by their spectral behaviour (light and colour) and their distribution in terrestrial space using specific measurements taken at a distance".

In addition, remote sensing is the result of the interaction between three fundamental elements: an energy source, a target and a sensor, and consists of measuring an electromagnetic signal emitted or reflected by a target. The source can be the sun or a satellite, depending on the type of energy to be captured (Amina, 2012).

Passive remote sensing relies on natural energy - sunlight reflected from the earth's surface.

Active remote sensing uses the energy emitted by satellites and reflected back to the satellites by the earth's surface.

### 2.7.2 Technical developments

Thanks to technical advances in remote sensing over the last few decades, earth observation data from sensors on board aircraft or spacecraft are increasingly diverse (monitoring plant resources, rainfall, the atmosphere, etc.) and rapidly available (Toukiloglou, 2007; Dubreuil, 2010; Diwakar *et al,* 2013). This remote sensing data is most often first available in image form. The interpretation of a satellite image is mainly based on the assumption that it is possible to recognise a certain number of features present on the earth's surface from the luminance values of these objects in the different windows of the electromagnetic spectrum. The data provided by earth observation satellites is defined above all by three main characteristics (Caloz and Collet, 2001).

- **spatial resolution:** this is the sensor's ability to distinguish between two objects that are close together. It is expressed in terms of pixel size.
- **temporal resolution:** this refers to the frequency with which images are acquired at a constant location. This is the period required for the satellite to revisit the same location (complete orbit cycle).

- **spectral resolution:** this is the smallest width of the spectral band recorded by the sensor. There are two modes of spectral resolution: panchromatic mode, comprising a wide band of spectral frequencies with low spectral resolution, and multispectral mode, comprising several bands of spectral frequencies with high resolution.

### 2.7.3 Data processing

A satellite image is made up of elementary components called "pixels", the value of which is defined by the mean radiance measurement. Most remote sensing software processes pixels one by one on the basis of their spectral properties (Bontemps, 2004).

The development of technologies has made it possible to obtain very high resolution images, which has led to a change in the way satellite data is processed. The number of mixed pixels, containing several objects or parts of objects, has been greatly reduced, making it much easier to interpret spectral signatures. These very high resolution images even call into question the relevance of the analysis scale, which is no longer necessarily the same as that of the pixel (Pekkarinen, 2002).

A new method has been developed to best process this new information: contextual classification or segmentation. It integrates contextual and spatial information. Pixels are still analysed on the basis of their spectral properties, but also on the basis of the relationships they have with their neighbours. The image is divided into several regions, the size of which is greater than that of the pixel, the grouping criteria being proximity, repeatability, surface area, shape and a certain homogeneity of texture (Lillesand and Kiefer, 1994; Pekkarinen, 2002). This approach could be very useful for processing very high resolution images.

To facilitate and optimise classification, it is important to select the most interesting spectral bands, i.e. those that are most discriminating. To do this, we need to know the spectral behaviour of the elements to be distinguished. The reflectance of bare soil, for example, is generally little affected by wavelength, but moisture content, organic matter concentration, percentage of iron and texture are all important.

Water, on the other hand, has a spectral signature with three absorption bands, 1.45µm, 1.9µm, 2.7µm; this can be seen in vegetation to a greater or lesser extent depending on the concentration of water in the leaf tissue (Bonn and Rochon, 1992). The spectral signature of vegetation shows two completely opposite behaviours in the visible range, particularly in the red, and in the PIR range, with the transition between the two resulting in a sudden increase in reflectance.

## 2.8 GEOGRAPHIC INFORMATION SYSTEM (GIS)

"A GIS is a set of spatially referenced data, structured in such a way that it is easy to extract summaries for use in decision-making" (Didier, 1990).

According to the Société française de photogrammétrie et de télédétection, "a GIS is a computer system that makes it possible, from various sources, to gather, organise, manage, analyse, combine, prepare and present geographically located information, contributing in particular to spatial management".

According to konecny (2003). "A GIS, in a narrow definition, is a computer system for the capture, manipulation, storage and visualisation of digital spatial data. In a broader definition, it is a digital system for the acquisition, management, analysis, modelling and visualisation of spatial data for the purposes of planning, administration and control of the natural environment and for socio-economic applications".

### 2.8.1 GIS functionalities

GIS are created to meet different demands. There are five (5) functionalities in GIS (Yacine, 2018).

- **abstraction:** designing a model that arranges data by geometric constituents and descriptive attributes that also allows relationships between objects to be established.
- **acquisition:** the software must have digitisation and data import functions.
- **archiving**: the software must have a large data storage capacity.
- **analysis:** ability to analyse geographical data (quantitative and statistical methods).
- **display:** ability to display geographic information in the form of maps, tables and graphs.

### 2.8.2 Ways of representing geographic information in a GIS

There are two ways of representing geographic information in a GIS (http://www.ign.fr).

**raster mode:** reality is broken down into a regular rectangular grid, organised into rows and columns, with each pixel in the grid having a grey scale or colour.

**vector mode:** the boundaries of spatial objects are described using their elementary components, i.e. points, arcs and polygons. Each spatial object is given an identifier that can be used to link it to an attribute table.

- **point:** The point represents any geographical feature whose scale dimensions cannot be taken into account.

- **arc (line):** represents any geographic feature whose length can be taken into account without its width.
- **polygon:** is any geographic feature whose dimensions can be considered in its representation.

# CHAPTER 3: MATERIAL AND DATA

## 3.1 EQUIPMENT

In this section we will discuss the instruments used to collect data in the field and the software used to carry out the various processing operations.

### 3.1.1 Instruments used

We used the following equipment to carry out our work. It consists of :

- a GARMIN etrex 10 GPS (Global Positioning System) for coordinates;
- a digital camera for taking photographs ;
- a notepad ;
- a pirogue and a lifejacket for river crossings;
- an ASUS CORE i3 64-bit computer for data processing.

### 3.1.2 Software or processing tools

The following software was also used:

- ENVI 5.3 for image processing ;
- ArcGIS 10.2 for map production ;
- Excel 2016 for statistical analysis and Word for word processing.

## 3.2 DATA

### 3.2.1 Satellite data

The satellite images used in this study are LANDSAT images (ETM+ and OLI). These images were downloaded from the USGS "Earthexplorer" web application (https://earthexplorer.usgs.gov) in the UTM Zone 29 projection system. Three images per study year were used to produce the various maps of aquatic plant cover between 2000 and 2018.

**Table I:** input data.

| Images | Scenes (Path/Row | Acquisition date |
|---|---|---|
| LANDSAT Enhance Thematic Mapper (ETM+) 2000 | 196-56 | 17/03/2000 |
| | | 05/06/2000 |
| | | 13/09/2000 |
| LANDSAT Enhance Thematic Mapper (ETM+) 2006 | 196-56 | 11/03/2006 |
| | | 22/06/2006 |
| | | 01/09/2006 |
| LANDSAT Enhance Thematic Mapper (ETM+) 2012 | 196-56 | 08/03/2012 |
| | | 15/06/2012 |
| | | 15/06/2012 |
| LANDSAT 8 Operational Land Image (OLI) 2018 | 196-56 | 19/03/2018 |
| | | 02/06/2018 |
| | | 02/09/2018 |

### 3.2.2 Map data

The cartographic data used in this study comes from the IGM (Institut Géographique du Mali) database. These are made up of vector data (boundaries of the Bamako area, roads and blocks).

## CHAPTER 4: METHODS

### 4.1 Satellite image acquisition

The dates chosen for image acquisition are very important in the study of landscape changes based on satellite data (Benvenuti, 1996; Girard and Girard, 1999; Oszwald *et al,* 2007; Daoudi *et al,* 2009).

The images were chosen according to the water level of the River Niger in the Bamako district. A curve of the elevation level (average daily coast) of the river at the Bamako station, obtained from the Direction Nationale d'Hydraulique in Bamako, was used to determine the average monthly flows of the River Niger.

However, the choice of image acquisition periods (March, June, September) was made in order to have a better knowledge of invasive aquatic plants during the low-water period of the river, which corresponds to March, during the high-water period (June) and when the river has reached its maximum level (September). We therefore used the elevation curve (average daily rise) of the river at the Bamako station to identify these periods.

### 4.2 Pre-processing of Landsat satellite images

The images have already been geometrically corrected by the USGS (U.S. Geological Survey) before publication. The pre-processing operations carried out were limited to radiometric and atmospheric corrections, as well as extraction of the study window.

- **Radiometric correction**

Radiometric correction is important for the image, as it enables the values of the pixels in the image (digital count) to be converted into reflectance values. The different bands in the images were therefore calibrated. This calibration consists of normalising the signal intensity in order to perform a multi-temporal analysis and compare the images. This task was carried out using ENVI 5.3 software.

- **Atmospheric correction**

This atmospheric correction reduces the effects of atmospheric behaviour (amount of water vapour, aerosol distribution) on electromagnetic radiation.

The method we used is the one that involves the sensor parameters at the time of shooting, called the FLAASH method, then the FLASH atmospheric correction tool in ENVI 5.1 allowed us to carry out this work.

- **Pan sharpening (image fusion)**

Image fusion is a widely used procedure for combining the fine resolution band of panchromatic images with the medium resolution multi-spectral bands of satellite images. The aim is to obtain a good perception of terrestrial detail (Yuhendra *et al.,* 2012).

The process is known as "Pan Sharpening". There are several different "Pan Sharpening" algorithms, depending on the degree to which they maximise image sharpness (Maurer, 2013).

However, we used the Gram-Schmidt technique, which is one of the most accurate. It is based on a spectral response function from the original sensor that estimates the characteristics of the input panchromatic band. More concretely, according to Maurer (2013), the operation is as follows.

- Simulation of the panchromatic band using multispectral bands

$$Pan_{sim} = \sum_{k=1}^{n} w_k \, MS_k$$

- **Equation 1:** Pan-Sharpening. Source: Maurer, (2013).

**$Pan_{Sim}$ :** Simulated panchromatic band; **$MS_k$ :** Multispectral band k ;

**Wk:** weighting coefficients **n:** number of multispectral bands.

**K € [1; n]:** multispectral band number.

- Apply the Gram-Schmidt transformation to the simulated panchromatic band and spectral bands, entering the simulated panchromatic band first.

The following equation gives the Gram-Schmidt transformation coefficients.

$$GS_{kl} = \frac{< MS'_k | MS_l >}{< MS'_k | MS'_k >} = \frac{\sum a_i^l < MS_i | MS_k >}{\sum \sum a_i^l a_j^l < MS_i | MS_j >}$$

**Equation 2:** Gram Schmidt transformation coefficient. Source: (Maurer, 2013)

(**GS:** Gram-Schmidt coefficient; **MS:** Multispectral band; **a**: weight vector)

- exchange of the panchromatic band with the first Gram-schmidt band.
- application of the inverse Gram-Schmidt transformation to form the new, enhanced spectral bands.

The Landsat ETM+ and OLI 2018 images each have panchromatic bands with a spatial resolution of 15m. We therefore grouped the bands together in the "Stacking". Their spatial resolution was therefore improved using the "Pan-Shaping" (Gram-Schmidt) algorithm.

➢ **Extraction from the study area**

Vector files of the Bamako area, obtained from the IGM (Institut Géographique du Mali) were used to delimit the study area. The data consisted of roads, blocks and the boundary of the District of Bamako. They were then applied to the image to extract the area.

### 4.3 Satellite image enhancement and extraction of land cover information

#### 4.3.1 Vegetation index

The Normalised Vegetation Index uses the Red (R) and Near Infra Red (PIR) channels. The different vegetation indices have the effect of increasing the contrast of the "chlorophyll vegetation" theme and reducing that of the other themes in the satellite image. The NDVI (Normalized Difference Vegetation Index) is one of the most commonly used indices. It varies from -1 to +1 and is calculated using the following formula:

$$\boldsymbol{NDVI = \frac{PIR - R}{PIR + R}}$$

In this study, the NDVI was applied to the Landsat bands in an attempt to isolate the rafts of invasive aquatic vegetation on the river. Since the aim was to discriminate between aquatic plants in the riverbed, this initial approach was supplemented by other treatments: colour composition and principal component analysis (PCA).

#### 4.3.2 Colour composition

The aim of this operation is to have a synthesis of information in order to make a good discrimination of the types of formation that cover the area. If the NDVI has enabled a separation between vegetation and non-vegetation occupation, then the aim here is to highlight vegetation sub-classes. However, the aim here is to highlight vegetation sub-classes. Each band of LANDSAT satellite imagery contains information about the study area. For this study, the following colour compositions (RGB) were used:

- For the ETM+ sensor, the combination of false colour bands is 4-3-2; for the years 2000, 2006 and 2012.
- For the 2018 OLI sensor the false colour band combination is 5-4-3.

Figure 7 shows the colour composition of the two sensors.

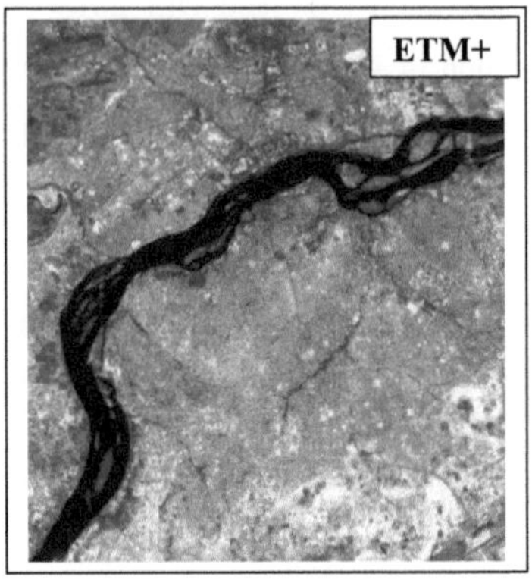


**Figure 7: Overview of "false colour" colour compositions of LANDSAT ETM+ and OLI images (bands 4-3-2 for ETM+ and bands 5-4-3 for OLI).**

### 4.3.3 Principal component analysis (PCA)

Although the colour compositions on the raw Landsat ETM+ (4-3-2) and OLI (5-4-3) bands proved to be more discriminating than the index, another approach was used to further improve the discrimination of invasive aquatic vegetation in the riverbed compared with the first two approaches. The Landsat visible and PIR (Near Infra-Red) spectral bands contain redundant information. The PCA applied to the images was therefore designed to improve the visual quality of the image by decorrelating the spectral bands and concentrating the information on the first three neo-channels formed. The colour composition produced from these neo-channels proved to be more discriminating for bare soil habitat, water, aquatic vegetation and terrestrial vegetation.

### 4.4 Data collection mission

In the field, the aim was to highlight invasive aquatic plants in the bed of the River Niger in the District of Bamako. The aim was to identify permanent infestation sites in order to describe plant formations, particularly aquatic plants. This description was possible thanks to a visual assessment carried out on the various permanent occupation stations. The basis for this mission was the set of GPS points selected from the images after enhancement (total of 10 points). Both banks of the river were covered, as well as the space between the Fahd Bridge and the Martyrs Bridge. In these various locations, the rafts of invasive aquatic vegetation were observed, their average diameters assessed and their GPS positions recorded. Finally, photos were taken at each site for image classification and to validate the aquatic plant class. A total of 210 points were visited during the field mission.

## 4.5 Image classification and mapping of the spatial distribution of invasive aquatic plants

### 4.5.1 Selection of training plots

This is the learning phase, which consists of establishing classification rules based on the knowledge available a priori (Nabil, 2000). In order to better identify the different land cover classes, the training plots were digitised on the coloured composition from the PCA neo-channels. The training plots were selected in such a way as to obtain the widest possible range of pixels representing the aquatic vegetation class, while taking care not to confuse them with terrestrial vegetation. A total of 70 points were selected, including 20 for invasive aquatic vegetation, for image classification.

### 4.5.2 C hoice of algorithm and implementation of classification

Supervised classification is a method of digital image interpretation based on the affiliation of pixels to thematic classes defined and recognised by the operator on the basis of knowledge of the terrain (Lacombe & Sheeren, 2007). In this method, the number of possible classes to be defined and the spatial location of each class are known from the terrain information. Several methods based on this approach have been developed, such as the connectionist approach based on the neural network method.

Based on our proven knowledge of the terrain, we used the "maximum likelihood" algorithm following the probabilistic approach in our image classification. This algorithm classifies unknown pixels by calculating, for each class, the probability of the pixel falling into the class with the highest probability.

However, if this probability does not reach the expected threshold, the pixel is classified as unknown. According to the study by Skupinski *et al* (2009), Shalaby and Tateish R, (2007). This algorithm uses the statistics of the training plot samples to calculate the probability of each pixel belonging to one of the land use classes with the highest probability.

### 4.5.3 Post classification

#### 4.5.3.1 Validation of the classification

Classification is validated by analysing a confusion matrix. In this study, a total of 140 points were used in the validation phase to produce the confusion matrix. These were visited sites that were not used for classification.

According to Toko (2014), the confusion matrix is a double entry table commonly referred to as a contingency matrix. The different reference classes used are positioned in the columns and the classes used during classification are included in the rows. This confusion matrix generates several indices (Bontemps, 2004):

***Overall accuracy:*** Number of well-ranked objects in relation to the total number of objects surveyed.

***The Kappa coefficient:*** characterises the ratio of well-ranked pixels to the total number of pixels surveyed (Skupinski *et al.,* 2009).

The Kappa index provides information on the agreement between the data to be classified and the reference data (Congalton, 1991).

#### 4.5.3.2 Digitisation and map rendering

Digitisation is the final stage in satellite image processing. This involves converting our classification into vector mode (polygons).

Cartographic rendering consists of creating the land use map by adding the semiological elements: the geographic North, the legend, the scale and the geographic coordinates.

### 4.6 Colonisation or occupation dynamics of invasive aquatic plants in the bed of the River Niger

Studies on land use change are of great importance, as they provide insight into current trends in environmental degradation processes (Khalid *et al.,* 2016).

To assess the dynamics of the occupation of a phenomenon, it is necessary to calculate the average annual rate of spatial expansion (T). In this study, we used the formula of Bernier (1992) cited by Oloukoi *et al* (2006), where the variable considered is the surface area (S).

The formula for calculating this index is :

T = [**S2-S1/t** - 1] x 100

Hence

T= rate of change (%)

S1 = surface area of the class at date t1

S2 = surface area of the class at date t2 (t2 sup t1)

*t* = number of years between the two dates.

### 4.7 Assessing the impact of aquatic plant colonisation dynamics on river flow

This calculation represents the water retention capacity for each land use class by calculating the **Cr** index. The calculation formula is :

$$Cr = \sum Pi * ai$$

Where **Pi** is the surface area of the study area occupied by land use class **i**, **ai** is a weighting coefficient representing the effective water retention capacity. Table II defines the **ai** with the land use classes. The **ai** have been defined for each class between 0 (zero retention) and 2 (maximum retention). This definition means that the **Cr** coefficient is between 0 and 200. If **Cr** is greater than 100, this means that the basin is still wild and has significant vegetation cover, and is not very conducive to increased runoff. On the other hand, if it is less than 100, then the basin is very degraded and conducive to significant runoff (Rachid *et al.,* 2016).

**Table II:** Water retention capacity coefficient for each of the occupancy classes.

| Occupancy classes | ai |
|---|---|
| Bare soil habitat | 0 |
| Other vegetation | 1 |
| Aquatic plants | 2 |

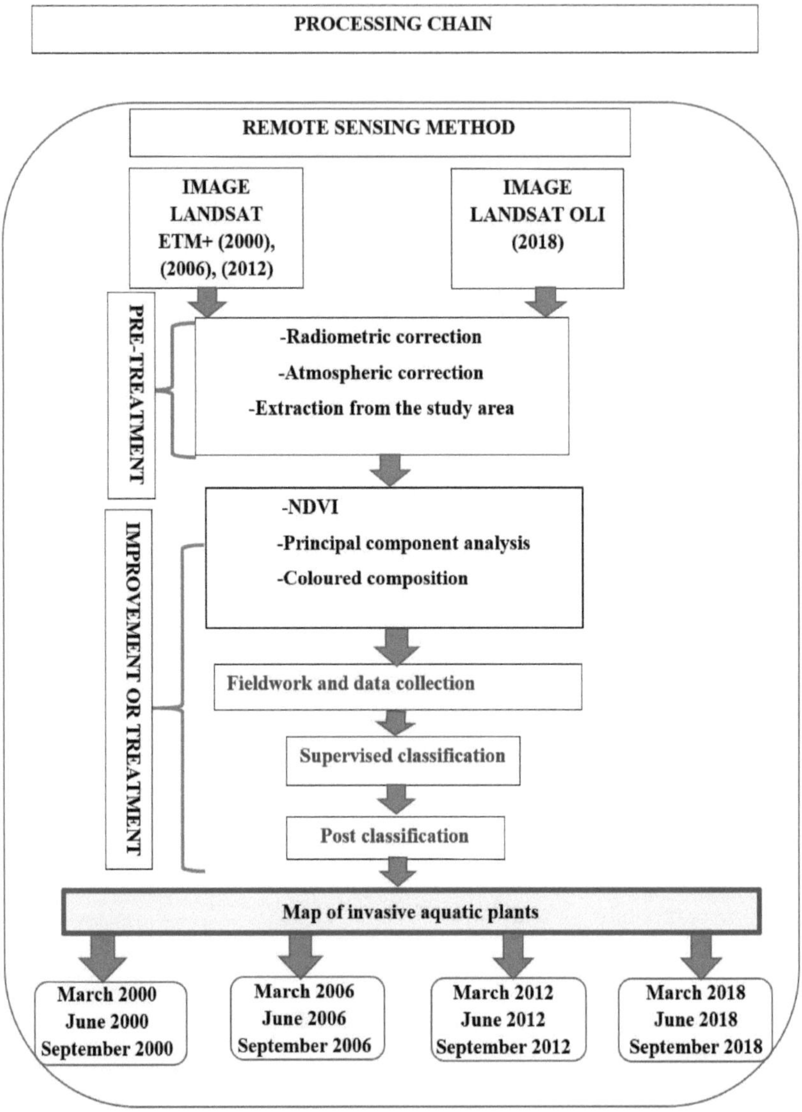


**Figure 8: Satellite image processing method**

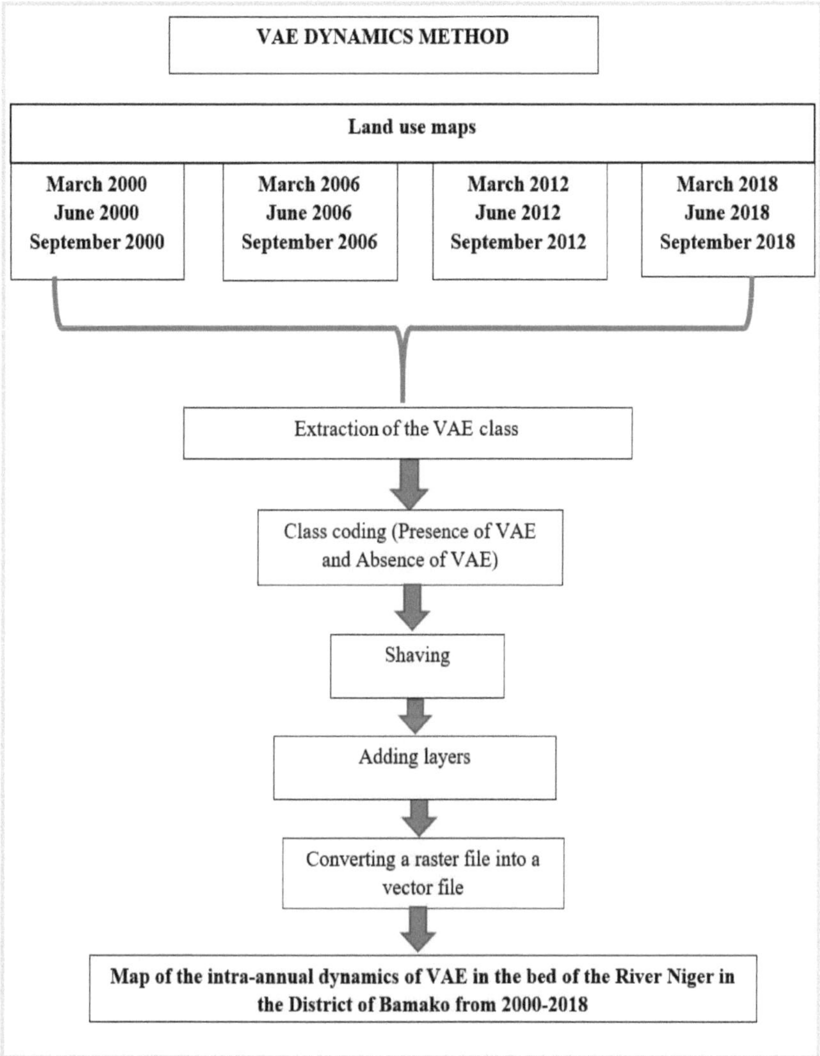


**Figure 9: VAE dynamics processing method**

## Partial conclusion

Remote sensing is a tool that is increasingly used not only in science but also in other fields. It offers users a wide field of study. Image pre-processing and processing or enhancement is applied systematically to any image data. Image classification is carried out to extract information about the area to be studied. Remote sensing is essential for detecting changes between images of different dates. It was therefore used in this study to help manage the colonisation dynamics of invasive aquatic plants in the bed of the River Niger. On this basis, the methodology used made it possible to draw up a map of the occupation of invasive aquatic plants over the study period.

## CHAPTER 5: RESULTS

### 5.1 Invasive aquatic plant cover maps for 2000, 2006, 2012 and 2018

#### 5.1.1 Types of land use in the study area

The Bamako area is well marked by different land use units. Given the purpose of this study, we have summarised the types of occupation into four (4) classes. Class 1 represents (bare soil habitat), class 2 (invasive aquatic vegetation), class 3 (the open water body of the River Niger) and class 4 (other vegetation).

1- **Housing (built) - bare ground**

Bare soil habitat includes all human constructions and undeveloped areas devoid of vegetation. Figure 10 illustrates the bare soil habitat in the study area.

**GPS coordinates :** X= -7, 99658 ; Y= 12,63139

**Figure 10: Habitat (buildings)-bare ground in the study area**

2- **Aquatic plants**

These are invasive aquatic plants in the bed of the River Niger. Figure 11 shows a permanent site of aquatic plants. This point is made up solely of freshwater hyacinth, which covers 80 to 100% of the water body.

**GPS coordinates :** X= -7, 98935 ; Y= 12, 62648

**Figure 11: Invasive aquatic plants in the River Niger water body**

### 3- The open water of the river

This is the River Niger in the Bamako district. Figure 12 shows the River Niger.

**GPS coordinates** : X= -7, 96518 ; Y= 12,63733

**Figure 12: Surface area of open water in the River Niger**

### 4- Other vegetation

This includes all woody and herbaceous vegetation, including market gardening on the banks of the River Niger. Figure 13 shows other vegetation in the study area.

**Figure 13: Other vegetation**

### 5.1.2 Evaluation of supervised classification

The confusion matrices were used to evaluate and validate the classifications (Tables III, IV, V and VI). Analysis of these classifications shows good overall accuracy for the entire study area. The first three images from the ETM+ sensor, i.e. the image from (2000; 2006 and 2012), show the following accuracies:

- The image from (2000) gives an overall accuracy of 99.07% for March, 89.50% for June and 90.64% for September, with a Kappa index of 0.98, 0.84 and 0.83 respectively.
- Similarly, the overall accuracy of the (2006) image is 92.32% for March, 90.86% for June and 95.14% for September, with a Kappa index of 0.86, 0.87 and 0.93 respectively.
- For the year 2012, the overall accuracy was 98.41% for March, 91.71 for June and 91.96 for September, with Kappa values of 0.97, 0.86 and 0.88 respectively.
- The last image in our study from the OLI sensor (2018) shows an overall accuracy of 97.17% for the month of March, 97.99 for the month of June and 98.06 for the month of September, with Kappa values of 0.95, 0.97 and 0.97 respectively. A more detailed analysis of these classes shows that the year classes (2000) have pixels that are better classified than the other dates.

Generally speaking, there is confusion between certain classes, such as the invasive aquatic plants class and the "Other Vegetation" class, and also between open water and the invasive aquatic plants class. There is also confusion between the "Bare ground" class and the "Other vegetation" class. However, a much more in-depth observation is that in Table III, in September, the "Bare ground" class is confused with the "Other vegetation" class by a high value.

**Table III:** Confusion matrix for the classification of the 2000 image

| Month | Classes | V.A | B.S | Water | To Vegeta |
|---|---|---|---|---|---|
| March | **V.A** | **97,28** | 0,00 | 0,07 | 0,50 |
| | **B.S** | 0,40 | **99,25** | 0,00 | 0,85 |
| | **Water** | 0,00 | 0,00 | **99,93** | 0,00 |
| | **To Vegeta** | 2,32 | 0,75 | 0,00 | **98,64** |
| | **Total** | 100 | 100 | 100 | 100 |

**SHARE CAPITAL:** 99.07% ;

**Kappa coefficient:** 0.98

| Month | Classes | V.A | B.S | Water | To Vegeta |
|---|---|---|---|---|---|
| June | **V.A** | **87,30** | 0,84 | 1,03 | 6,86 |
| | **B.S** | 0,75 | **91,63** | 0,00 | 5,14 |
| | **Water** | 3,74 | 0,00 | **98,97** | 0,00 |
| | **To Vegeta** | 8,21 | 7,53 | 0,00 | **88,00** |
| | **Total** | 100 | 100 | 100 | 100 |

**P L:** 90.64

**kappa coefficient:** 0.84

| Month | Classes | V.A | B.S | Water | To Vegeta |
|---|---|---|---|---|---|
| September | **V.A** | **79,30** | 1,09 | 0,11 | 5,73 |
| | **B.S** | 0,00 | **83,04** | 0,00 | 0,00 |
| | **Water** | 1,00 | 0,22 | **99,89** | 9,17 |
| | **To Vegeta** | 20,69 | 15,65 | 0,00 | **85,09** |
| | **Total** | 100 | 100 | 100 | 100 |

**P L:** 89.50%.

**kappa coefficient:** 0.83

**With :**

**V A E=** Invasive aquatic vegetation; **B S=** Bare soil; **Open water** (river); **A V=** Other vegetation.

**Table IV:** Confusion matrix for the classification of the 2006 image

| Month | Classes | V.A | B.S | Water | To Vegeta |
|---|---|---|---|---|---|
| | V.A | **94,42** | 0,04 | 3,94 | 2,52 |
| | B.S | 0,26 | **92,16** | 1,46 | 6,48 |
| March | Water | 0,09 | 0,00 | **93,52** | 0,00 |
| | To Vegeta | 5,23 | 7,80 | 1,08 | **91,00** |
| | Total | 100 | 100 | 100 | 100 |

**P L:** 92.32

**Kappa coefficient:** 0.86

| Month | Classes | V.A | B.S | Water | To Vegeta |
|---|---|---|---|---|---|
| | V.A | **86,56** | 0,19 | 0,00 | 8,49 |
| | B.S | 1,29 | **93,12** | 0,00 | 3,86 |
| June | Water | 4,13 | 0,00 | **100** | 0,00 |
| | To Vegeta | 8,01 | 6,69 | 0,00 | **87,64** |
| | Total | 100 | 100 | 100 | 100 |

**P L:** 90.86

**kappa coefficient:** 0.87

| Month | Classes | V.A | B.S | Water | To Vegeta |
|---|---|---|---|---|---|
| | V.A | **89,91** | 0,00 | 2,78 | 4,69 |
| | B.S | 0,92 | **97,57** | 0,00 | 0,78 |
| September | Water | 1,83 | 0,00 | **97,22** | 0,00 |
| | To Vegeta | 7,34 | 2,43 | 0,00 | **94,53** |
| | Total | 100 | 100 | 100 | 100 |

**P L:** 95.14

**kappa coefficient:** 0.93

**With :**

**V A E**= Invasive aquatic vegetation; **B S**= Bare soil; **Open water** (river); **A V**= Other vegetation.

**Table V:** Confusion matrix for the classification of the 2012 image

| Month | Classes | V.A | B.S | Water | To Vegeta |
|---|---|---|---|---|---|
| March | **V.A** | **98,84** | 0,00 | 0,00 | 0,78 |
| | **B.S** | 0,00 | **98,16** | 0,00 | 3,36 |
| | **Water** | 0,11 | 0,00 | **100** | 0,00 |
| | **To Vegeta** | 1,05 | 1,84 | 0,00 | **95,87** |
| | **Total** | 100 | 100 | 100 | 100 |

**P L:** 98.41

**Kappa coefficient:** 0.97

| Month | Classes | V.A | B.S | Water | To Vegeta |
|---|---|---|---|---|---|
| June | **V.A** | **92,71** | 0,10 | 3,17 | 4,46 |
| | **B.S** | 0,65 | **92,58** | 0,40 | 8,36 |
| | **Water** | 2,73 | 0,00 | **96,43** | 0,00 |
| | **To Vegeta** | 3,91 | 7,32 | 0,00 | **87,31** |
| | **Total** | 100 | 100 | 100 | 100 |

**P L:** 91.96

**Kappa coefficient:** 0.86

| Month | Classes | V.A | To Vegeta | Water | B.S |
|---|---|---|---|---|---|
| September | **V.A** | **68,77** | 0,73 | 4,90 | 0,00 |
| | **To Vegeta** | 16,60 | **90,10** | 1,36 | 2,44 |
| | **Water** | 13,44 | 0,00 | **92,23** | 0,00 |
| | **B.S** | 1,19 | 9,17 | 1,50 | **97,54** |
| | **Total** | 100 | 100 | 100 | 100 |

**P L:** 91.73

**Kappa coefficient:** 0.88

**With :**

**V A E**= Invasive aquatic vegetation; **B S**= Bare soil; **Open water** (river); **A V**= Other vegetation.

**Table VI:** Confusion matrix for the classification of the 2018 image

| Month | Classes | V.A | B.S | Water | To Vegeta |
|---|---|---|---|---|---|
| March | V.A | **94,00** | 0,00 | 5,60 | 1,10 |
| | B.S | 0,24 | **98,93** | 0,00 | 1,17 |
| | Water | 4,26 | 0,00 | **94,40** | 0,00 |
| | To Vegeta | 1,51 | 1,07 | 0,00 | **97,73** |
| | Total | 100 | 100 | 100 | 100 |

**P L:** 97.17

**Kappa coefficient:** 0.95

| Month | Classes | V.A | B.S | Water | To Vegeta |
|---|---|---|---|---|---|
| June | V.A | **97,33** | 0,00 | 0,55 | 5,69 |
| | B.S | 0,00 | **99,76** | 0,55 | 0,00 |
| | Water | 1,07 | 0,00 | **98,90** | 0,00 |
| | To Vegeta | 1,60 | 0,24 | 0,00 | **94,31** |
| | Total | 100 | 100 | 100 | 100 |

**P L:** 97.99

**kappa coefficient:** 0.97

| Month | Classes | V.A | B.S | Water | To Vegeta |
|---|---|---|---|---|---|
| September | V.A | **99,07** | 0,00 | 1,47 | 0,00 |
| | B.S | 0,00 | **97,79** | 0,00 | 2,97 |
| | Water | 0,00 | 0,00 | **98,53** | 0,00 |
| | To Vegeta | 0,93 | 2,21 | 0,00 | **97,03** |
| | Total | 100 | 100 | 100 | 100 |

**P L:** 98.06

**kappa coefficient:** 0.97

**With :**

**V A E**= Invasive aquatic vegetation; **B S**= Bare soil; **Open water** (river); **A V**= Other vegetation.

### 5.1.3 Mapping invasive aquatic plants in the bed of the River Niger in the Bamako district

#### 5.1.3.1 Mapping invasive aquatic plants in 2000

In 2000, aquatic plants occupied a large surface area in the bed of the River Niger at Bamako (Figure 15). Over the course of the same year, there was a change in invasive aquatic plants between the months studied (Figure 14). On the various maps, we can see a high concentration of aquatic plants, especially at the mouths of the city's tributaries and wastewater drainage channels (circled areas).

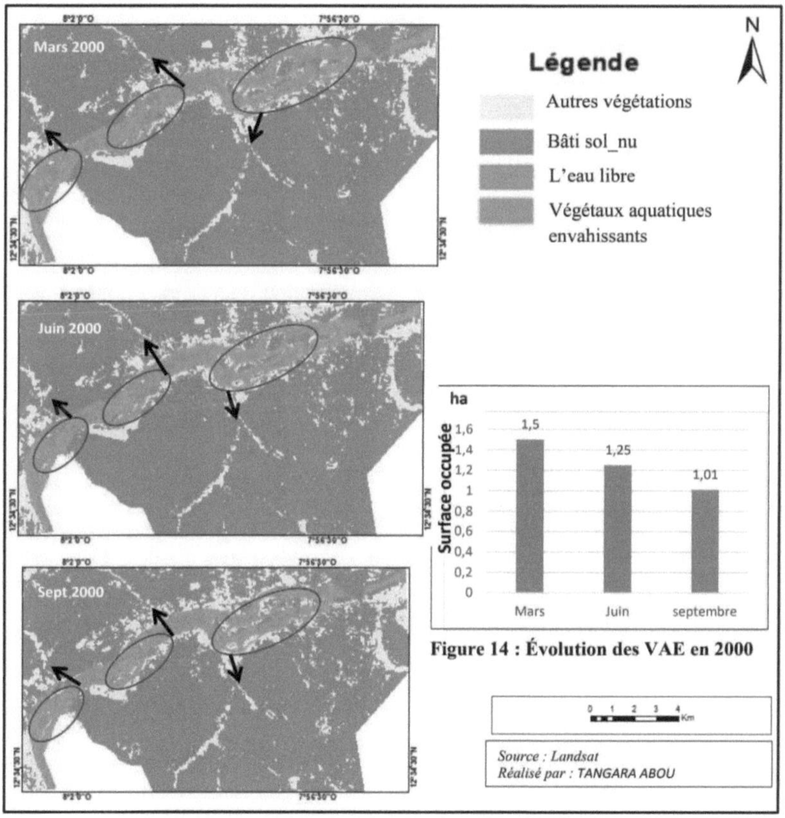


**Figure 14 : Évolution des VAE en 2000**

**Figure 15: Map of aquatic plant cover in 2000**

### 5.1.3.2 Mapping invasive aquatic plants in 2006

In 2006, the evolution of aquatic vegetation remained almost equal in terms of surface area, with a total of 2.55 hectares. Divided between the months studied as follows: March 0.92 ha, June 0.88 ha and September 0.75 ha (Figure 16). The areas circled on the maps show more encroachment by invasive aquatic plants (Figure 17).

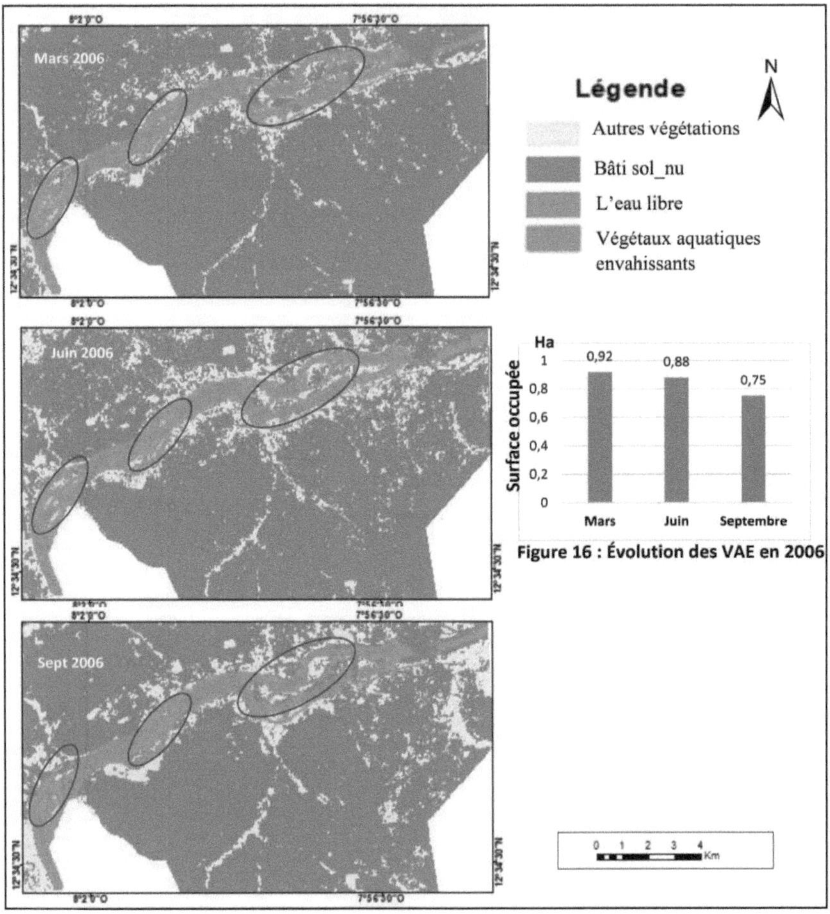


**Figure 17: Map of proliferation of invasive aquatic plants in 2006**

### 5.1.3.3 2012 mapping of invasive aquatic plants s

Figure 18 shows changes in aquatic vegetation between the study months, with September having the smallest surface area at 0.89 hectares. The densest areas are circled on the maps (Figure 19).

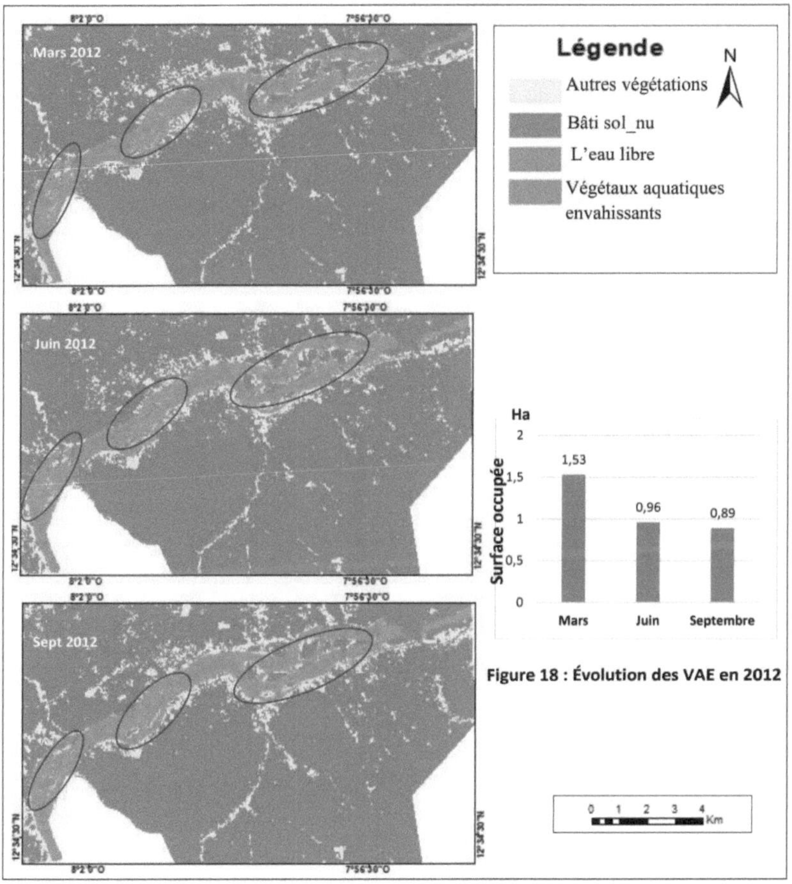


**Figure 19: Map of aquatic plant cover in 2012**

### 5.1.3.4 Mapping invasive aquatic plants in 2018

In 2018, aquatic vegetation varied between the study months, with the maximum in March at 2.02 hectares and the minimum in September at 1.21 hectares (Figure 20). The most concentrated areas are always circled on the maps (figure 21). These areas are heavily colonised by aquatic plants in the river bed.

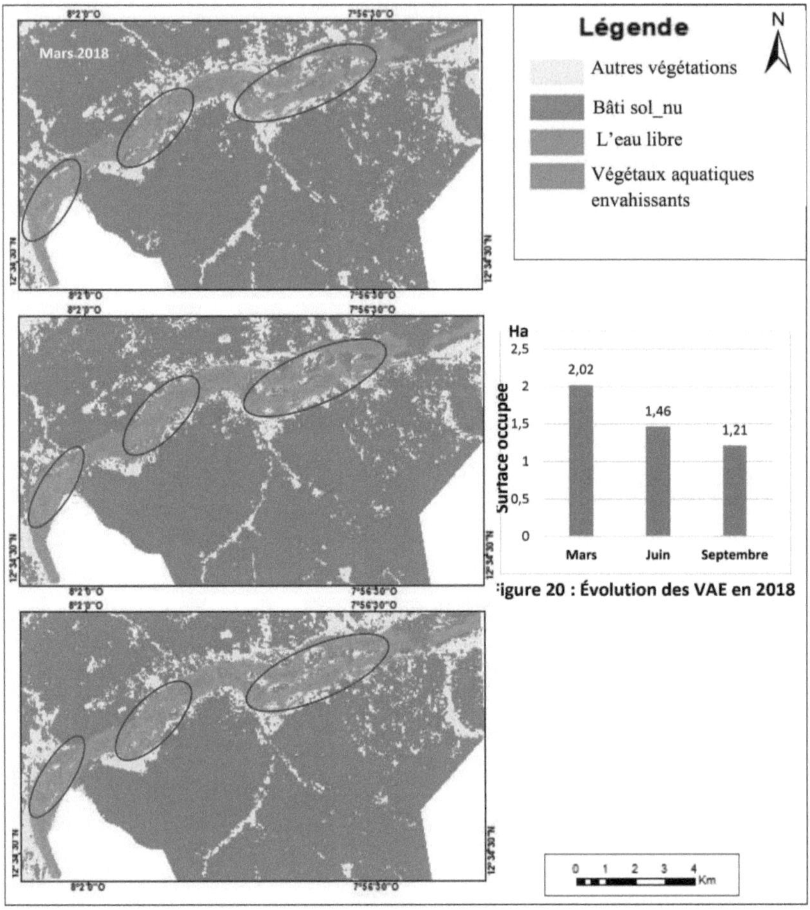


**Figure 21: Map of aquatic plant cover in 2018**

### 5.1.4 Analysis of the colonisation dynamics of invasive aquatic plants in the riverbed from 2000 to 2018

#### 5.1.4.1 Map of monthly colonisation dynamics of invasive aquatic plants

In the analysis of the monthly dynamics of invasive aquatic plants, it was found that over all the months studied, the areas occupied by aquatic plants were unevenly distributed (Figure 22 below).

Initially, in March, there were more stable areas of invasive aquatic plants than areas that were disappearing. In the same month, we also noted the presence of new appearances of invasive aquatic vegetation (IAV). Unlike the month of March, in June there was a slight domination of disappearing areas. Thus, the new appearances of IAS are tending to reach the level of the stable zones.

Finally, September saw more areas of instability for invasive aquatic plants. During this month, new appearances of EABs were observed and the stable zones of the previous months were transformed into zones of disappearance.

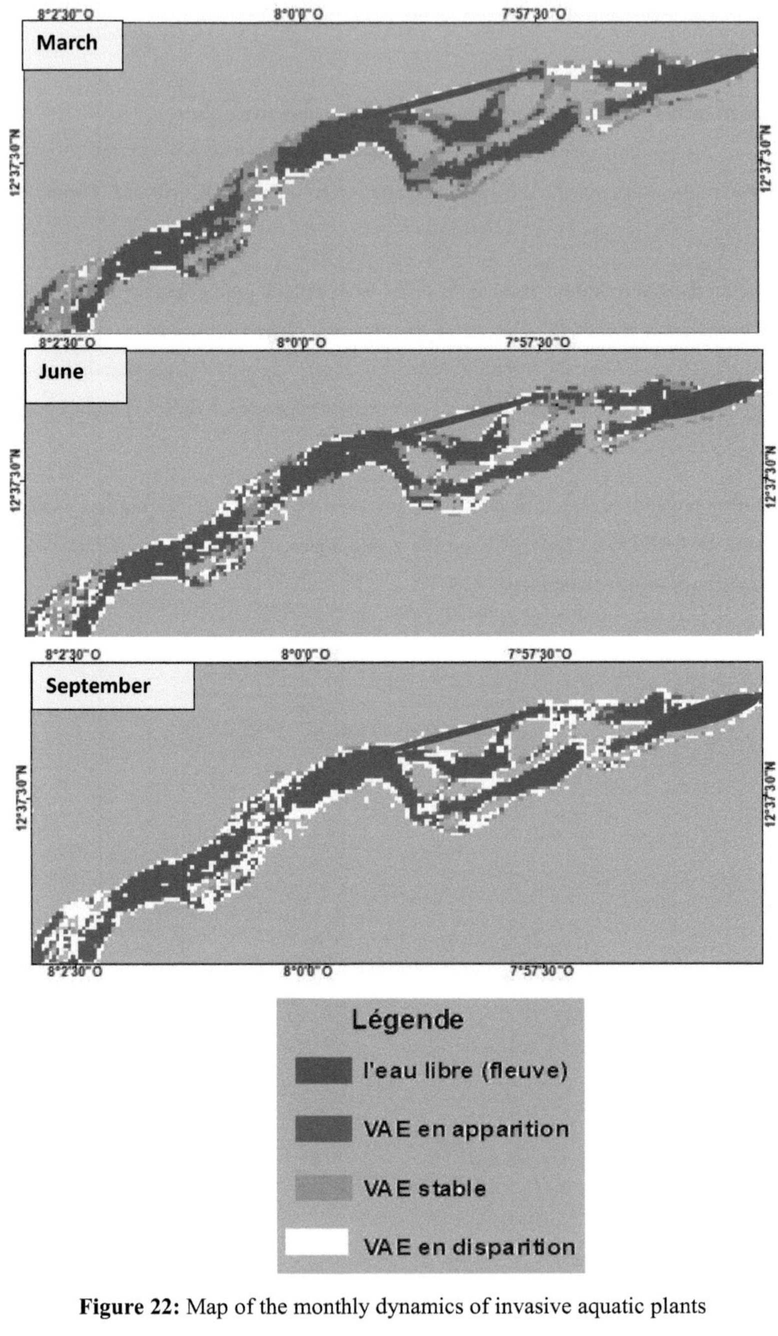


**Figure 22:** Map of the monthly dynamics of invasive aquatic plants

#### 5.1.4.2 Monthly spatio-temporal variation in the area occupied by invasive aquatic plants in the riverbed

Figure 23 below shows the monthly variation in invasive aquatic vegetation and the mean monthly discharge from the river. On this curve, we can see that the invasive aquatic vegetation on the river is depressed from June onwards. On the other hand, the month of June corresponds to the change in the mean monthly flow of the river in the study area. The following graphs show the variation in aquatic plants (dotted line) on the river from 2000 to 2018, and the change in the mean monthly flow (solid line) of the river from 2000 to 2018.

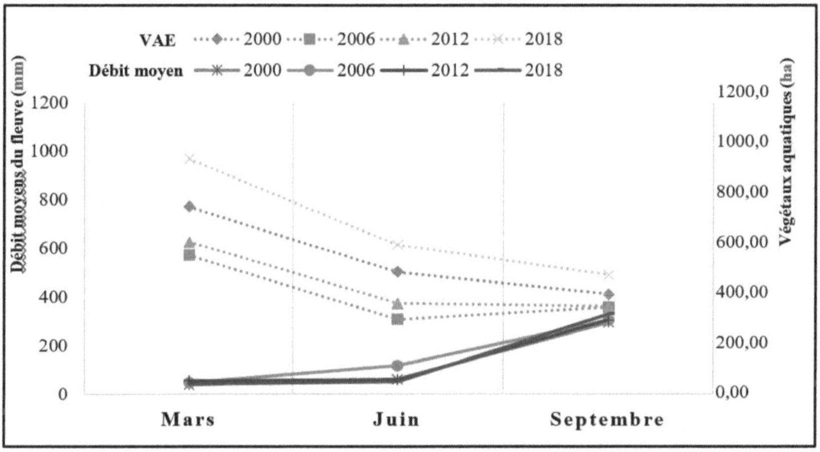


**Figure 23:** Monthly variation in invasive aquatic plants and mean river flow

### 5.1.4.3 Annual colonisation dynamics of invasive aquatic plants

Over all the years studied, the dynamics of aquatic plants varied from one year to the next (Figure 24 below). The different study years show a spatial discontinuity in invasive aquatic plants (IAV). The maps also show years in which the presence of invasive aquatic plants is high and years in which the presence of invasive aquatic plants is low.

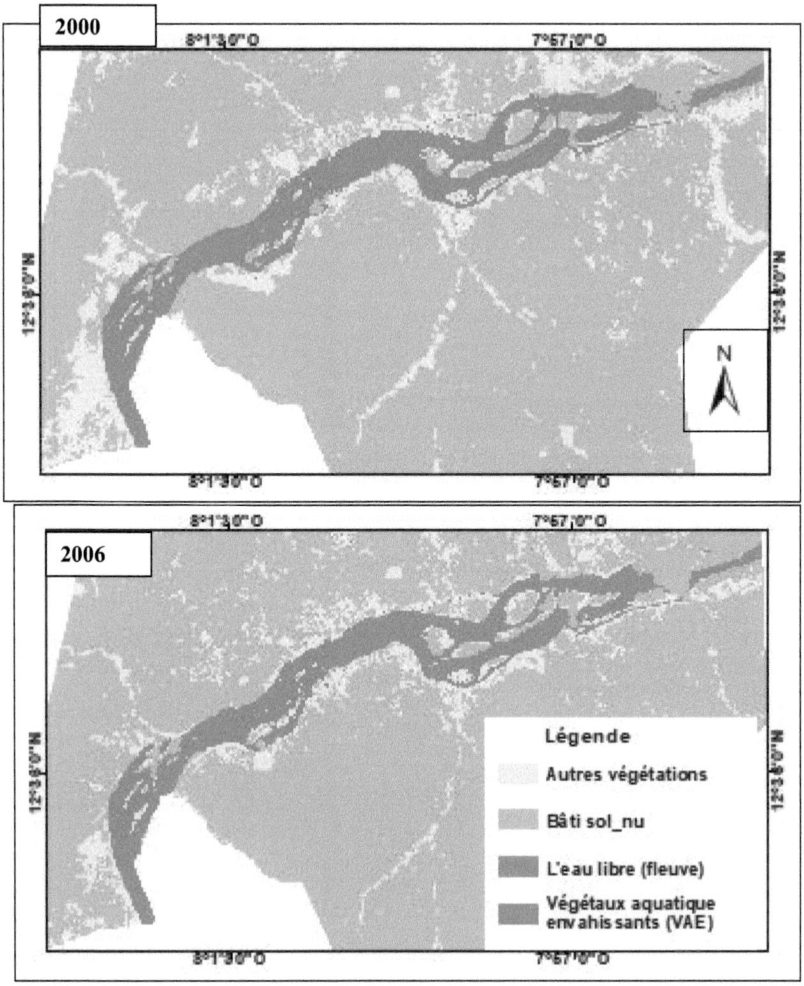


**Figure 24:** Annual dynamics of invasive aquatic plants in the riverbed

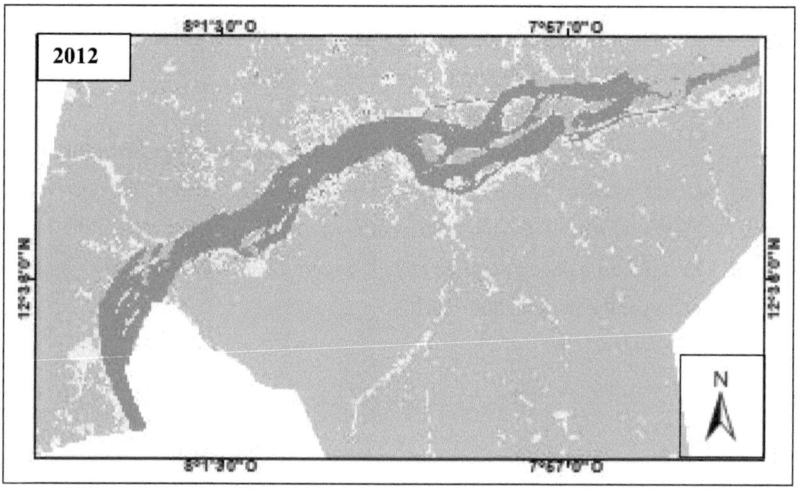


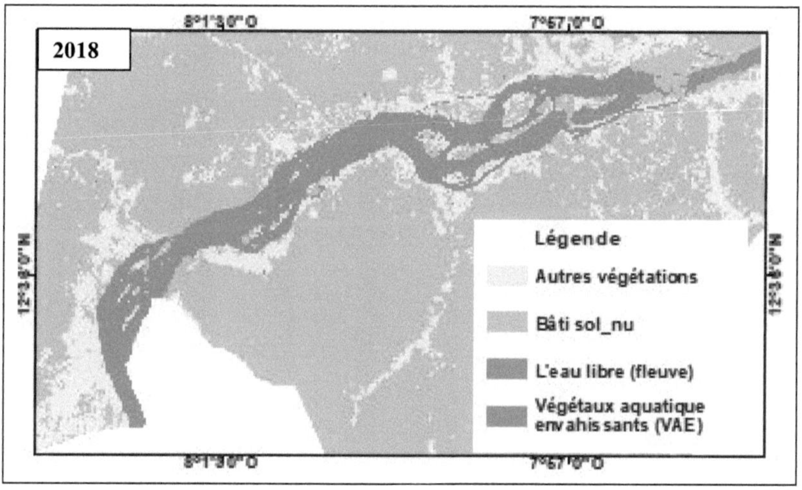


**Figure 24:** Annual dynamics of invasive aquatic plants in the riverbed

#### 5.1.4.4 Annual spatio-temporal trends in invasive aquatic plants from 2000 to 2018

Between 2000 and 2018, there was an increase in the area colonised by invasive aquatic plants in the riverbed. Although between 2000 and 2006 there was a slight decrease in the areas colonised, from 2006 onwards, the areas colonised by invasive aquatic plants have continued to increase. This increase was observed for both flood and deflooding periods. The low-water period (March) is a favourable time for the proliferation of AEVs, whereas during the high-water period, the areas colonised decrease, often by half. Figure 25 shows changes in colonised areas over the year and over the period 2000 - 2018.

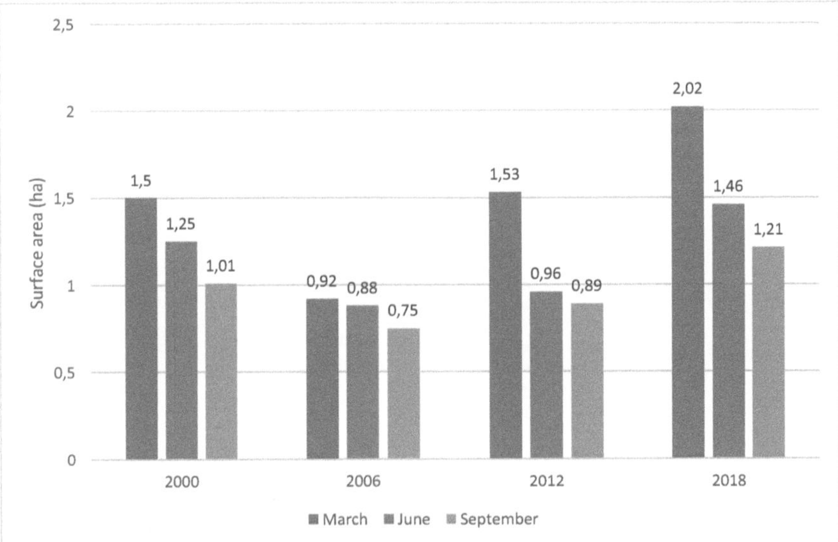


**Figure 25:** Annual temporal trends in invasive aquatic plants from 2000-2018

### 5.2 Average annual rate of spatial expansion (T) from 2000 to 2018

The average rate of spatial expansion was assessed over the study period using the Bernier (1992) formula.

T = [**S2-S1/t** - 1] x 100 Where T= rate of change (%)

S1 = surface area of the class at date t1

S2 = surface area of the class at date t2 (t2 sup t1)

*t* = number of years between the two dates.

After the assessment, we obtain the spatial expansion rates in the riverbed from 2000 to 2018, as shown in Table VII.

**Table VII:** Average annual rate of expansion (T) from 2000 to 2018

| **Years** | **2000-2006** | **2006-2012** | **2012-2018** | 2000-2018 |
|---|---|---|---|---|
| **Invasive aquatic plants (IAV)** | -20,16% | 13,83% | 21,83% | 15,50% |

Over all the years studied, the average rate of expansion (T) was 15.50%. In addition, a -20.16% decline in invasive aquatic plants was observed between 2000 and 2006.

**5.3 Impact of the proliferation of invasive aquatic plants on the flow of the river**

The impact of the proliferation of aquatic plants on the flow of water in the river was quantified by the retention coefficients (Cr) as shown in Table VIII.

**Table VIII:** Changes in the water retention coefficient (Cr) on the river between 2000 and 2018

| Occupancy classes | **Cr 2000** | **Cr 2006** | **Cr 2012** | **Cr 2018** |
|---|---|---|---|---|
| Bare soil habitat | 0 | 0 | 0 | 0 |
| **Aquatic plants** | **7,52** | **5,10** | **3,06** | **9,38** |
| Other vegetation | 6,22 | 8,33 | 4,66 | 4,4 |

In this table, we can see that the **Cr** values in our occupation classes, specifically the class of aquatic plants that concerns us, are less than 100. However, we can say that from 2000 to 2018 the evolution of aquatic plants has not reached the average that could prevent the river from flowing. But it remains a phenomenon to be monitored, given its degree of proliferation in the riverbed in the District of Bamako.

## CHAPTER 6: DISCUSSION

### 6.1 Classification

The various confusion matrices resulting from the classification show pixels well classified at over 70% per class, with a higher proportion of classes having accuracies above 90%. These accuracies are acceptable because, according to Congalton (1991), a classification is deemed acceptable when the overall accuracy is close to 80%. According to Pontius (2000), when the Kappa index is between 50 and 75%, the classification adopted is valid and the results can be used judiciously.

In the present study, the Kappa indices are greater than 85%. It should be noted that the high cartographic accuracies obtained may also depend on the small number of classes used and the definition of homogeneous plots when choosing training sites (Caloz and Collet, 2001).

Confusion exists between certain classes, such as the aquatic plants class and the "Other Vegetation" class. This confusion is likely because market gardens on the river bank often give the same signatures as aquatic plants. This confusion can also be explained by the fact that, during the rainy season, the vegetation cover of the land is dense. Given the similarity in colour, they may have the same spectral signature. The confusion between "Bare ground" and "Other vegetation" can be explained by the fact that the market gardens on the river bank have been harvested and the space is increasingly taking on the form of bare ground. Another explanation is that the area has been cultivated but the crops have not yet started to grow.

### 6.2 Dynamics of aquatic plant cover in the riverbed in the Bamako district

The contribution of spatial remote sensing can be summarised by highlighting the general trend in land cover dynamics (Rachid *et al.,* 2016). In our specific case, this trend reveals an increase in invasive aquatic plants in the bed of the River Niger from 2000 to 2018, with a sum expansion of 15.50% during the study period. This significant increase in invasive aquatic plants in the study area is due to the city's wastewater. This wastewater reaches the river through untreated drains, encouraging the proliferation of invasive aquatic plants. Our result is comparable to that of Jean (2018), in his study of the intra- and inter-annual dynamics of the proliferation of invasive plants on Lake Tengréla in Burkina Faso. His study was made possible by MODIS satellites, of which he used 411 images, i.e. 23 images per year over the period from 2000 to 2017. According to the author, an annual increase of 60% in the area colonised by invasive plants on Lake Tengréla was observed between 2000 and 2017. This indicates rapid growth of aquatic plants in water bodies. This increase in the surface area of aquatic plants depends on several parameters (chemical and organic). Anthropogenic pollution makes available in water a large quantity of nutrients necessary for the proliferation of invasive aquatic plants (Mandi *et al.,* 1992). Other work already carried out on invasive aquatic plants is confirmed by our results, in particular the work of Blé (2008), over the period 2002-2005, which identified the profitability of fishing on Lake Taabo, which is partially covered by invasive aquatic plants. The increase in aquatic plants is estimated at 33.30% over the period 2002 to 2005.

In terms of monthly dynamics, the curve representing invasive aquatic plants in the riverbed shows a regression in aquatic plants as a function of the average flow of the river in the District of Bamako. The flow of the river varies logically with rainfall. This is why the river begins to flood in early June during the winter season, and reaches low water during the dry season. The month of June, which marks the start of the rainy season, also corresponds to the decline in invasive aquatic vegetation in the riverbed. This is because the rise in the river's water level, accompanied by strong run-off, loosens the rafts of invasive aquatic vegetation (IAV) from one point to another. This is how, in September, when the maximum amount of rain falls, the most invasive aquatic plants disappear. During the dry season in March, when the river is at low water and there is not much run-off, there is a lot of invasive aquatic vegetation on the river.

In terms of annual dynamics, AEVs decreased between 2000 and 2006. This reduction in the surface area of invasive aquatic plants on the river can be explained by the increase in the river's flow rate, according to the driving force behind the dynamics of EAB. But it is also due to human intervention, since between 2000 and 2006 control initiatives were envisaged by technical services in Mali and by local people. Most of these initiatives were based mainly on manual and biological control. It is therefore highly likely that these interventions had a positive effect on EAB, but that this effect did not last. This is why, from 2012 to 2018, we have seen an increase in invasive aquatic plants. This increase in the extent of EABs on the river could be explained by the river's pollution. It has been found that the areas where EABs are concentrated correspond to the few wastewater drainage channels in Bamako. Other studies already carried out confirm that waste from Bamako's industrial units, such as the refrigerated abattoir, sotamali, synplast and sipal, is a source of river pollution (Maïga, 2012). The discharge of organic and chemical matter into the river water leads to the enrichment of EABs and therefore to an increase in their colony in the river bed.

## GENERAL CONCLUSION AND RECOMMENDATIONS

Ultimately, this study highlights the contribution of remote sensing in monitoring the spatial dynamics of invasive aquatic plants in the bed of the River Niger in the District of Bamako. The study was made possible by Landsat ETM+ and OLI satellite data from 2000 to 2018. The results of the mapping of invasive aquatic plants indicate :

- variations in invasive aquatic plants on the river from one year to the next. During our study period, there were years when aquatic plants were expanding and years when they were decreasing. Nevertheless, our study revealed the extent of the presence of aquatic plants in the river. We also note that the decline in invasive aquatic plants is linked to rainfall. The curve representing aquatic vegetation and the average flow rate of the river showed a decrease in invasive aquatic vegetation from June onwards. This month marks the start of the rainy season in the study area, and as soon as the rainy season sets in, the area infested by invasive aquatic plants gradually decreases. In any case, we can say that rainfall plays a determining role in the instability of invasive aquatic plants on the river.
- Analysis of the map of the annual and monthly dynamics of aquatic plants over the study period showed stable zones, the appearance of new sites and zones that were disappearing. The various dynamic maps show a high concentration of invasive aquatic plants on the river.

Complete and regular mapping of this kind makes it possible to specify the nature of the evolution of invasive aquatic plants in the sites studied. Our study, carried out on the river in the District of Bamako, enabled us to analyse the colonisation dynamics of invasive aquatic plants over the 18-year study period, i.e. from 2000 to 2018.

It is therefore advisable to take measures to protect bodies of water in order to better manage invasive aquatic plants. To this end, we recommend :

- raising public awareness of the impact of water pollution on the proliferation of aquatic plants;
- enforcement of existing legislation on the discharge of waste into the River Niger;
- organising the continuous removal of invasive aquatic vegetation during low-water periods in the river;
- constant monitoring of the evolution of invasive aquatic plants to improve management.

- ➢ the structures responsible for safeguarding the River Niger in Mali to take account of the results of this study in the management of the VAE.

**OUTLOOK**

- ❖ carry out this study with a finer sensor to maximise the detection of smaller rafts of invasive aquatic vegetation (IAV) ;
- ❖ Produce a detailed map of land-use dynamics, to gain a better understanding of the sources of nutrients for VAE ;
- ❖ to extend this study to the whole of Mali in order to gain a better understanding of the areas where VAE proliferate in the bed of the River Niger.

## BIBLIOGRAPHICAL REFERENCES

**Niger River Basin Agency. (2004).** Study on the control of water hyacinth in the River Niger. Final report 25 p.

**Niger Basin Authority. (2002).** Mission, current activities and prospects, Information document, 9 p.

**Alhou B. (2009).** Impact des rejets de la ville de Niamey (Niger) sur la qualité des eaux du fleuve Niger, PhD thesis, Facultés Universitaires Notre-Dame de la Paix NAMUR (Belgium) and Faculté des Sciences département de biologie de l'université Abdou Moumouni (Niger), 299p.

**Amina B., (2012).** Use of GIS and remote sensing in the study of vegetation cover dynamics in the Oued Bouguedfine sub-watershed (wilaya of chlef). 12p.

**Barret S., Forno I.W. (1982).** Style morph distribution in new world populations of *Eichhornia crassipes* (Mart.) Solms-Laubach (water hyacinth), Aquatic. **N°13**, pp. 299-306.

**Benvenuti A. (1996).** Analyses de télédétection : projet d'évaluation des interventions de conservation et récupération de l'environnement *(*PEICRE*)*. Technical report, CeSIAUTA, Niamey, Niger. 119 p.

**Bernier B. (1992).** Introduction to macroeconomics. Dunod, Paris, 127p

**Blé Acca Kouakou Serge R. (2008).** Profitability of fishing on Lake Taabo partially covered by invasive aquatic plants: Revue de Géographie Tropicale et d'Environnement, N°2, pp.53-54.

**Bontemps S. (2004).** Cartographie et interprétation de l'évolution du développement territorial par télédétection spatiale au Cambodge, dissertation from the Faculty of Agronomic Sciences, UCL, Louvain-la-Neuve, 111p.

**Bonn F., Rochon G. (1992).** Précis de télédétection. Principe et méthodes, presses de l'université de Quebec, Canada.vol.1 ; 485p.

**Boulerie, (2008).** Introduction to remote sensing, based on the Olivier Joinvilles course at the ENSP. 5p.

**Cheickna D., Bouréma D., Amadou D. (2004).** L'étude d'identification des villages affectés par les plantes aquatiques nuisibles proliférantes dans le bassin du fleuve Niger. Final report 7p.

**ECOWAS. (2003**). Project on the integrated management of invasive aquatic plants. Summary of the social and environmental impact assessment (EIA) study, October 2003. 6p.

**ECOWAS. (1996).** Projet de lutte contre les végétaux aquatiques flottants dans les pays membres de la CEDEAO, vol.1. Main Report, April 1996. 46p.

**Cissé C., Lamine D., Almoustapha F. and Pierrick G. (2007).** Future of the River Niger. In Freshwater Contamination (Proceedings of Rabat Symposium S4, April-May 1997), IAHS Publ. N°27, pp. 314-315.

**Caloz R., Collet C. (2001).** Précis de télédétection, traitements numériques d'images de télédétection (French-speaking universities). Presses Polytechniques du Québec, Canada. Vol.3, pp.251-254.

**Castillon P. (2005).** "Le phosphore, sources flux et rôles pour la production végétale et l'eutrophisation". Production animales- Paris -Institut national de la recherche agronomique n°18, vol.3, 153p.

**Congalton R.G. (1991).** A review of assessing the accuracy of classifications of remotely sensed data. Remote Sensing of Environment, N°37, pp. 35-46.

**Daoudi M., Salmon M., Dewitte O., Gerard P., Abdellaoui A., Ozer A. (2009).** Prediction of gully erosion in Algeria: towards a new probabilistic approach using multisource data. Journées d'Animation Scientifique de l'AUF, 7p.

**Dagno K., Lahlahi R., Friel D., Bajji M., Juakli H. (2007).** The problem of the water hyacinth, *Eichhornia crassipes,* in tropical and subtropical areas of the world, in particular its eradication by biological control using phytopathogens, Vol. 11, N°4, pp.299-311.

**Dembélé B. (1999).** Mise au point des techniques de lutte intégrée contre la jacinthe d'eau et d'autres plantes aquatiques nuisibles 12p.

**Dembélé B. (1994).** Water hyacinth, a scourge for Mali's watercourses? Mali. CILSS, Sahel PV info. 63, 8 p.

**Dembélé B., Diarra A., Diarra C., Sarra S., Traore M.N. (2002).** Le problème de la jacinthe d'eau Mali, état des recherches. In: Gestion des Ressources et Aménagement du Fleuve Niger. Des connaissances scientifiques pour la décision publique. IRD Edition Paris, 2002, pp. 57-62.

**Diarra B. (2003).** Structure urbaine et dynamique spatiale à Bamako- Mali, Donniya, Bamako, 163 p.

**Diarra B. (1999).** Dynamique spatiale et politiques urbaines à Bamako: le rôle des images-satellite SPOT dans la gestion des villes. Marseille (Phd), Université Aix-Marseille1, Marseille1, 245p.

**Didier M. (1990).** Utilité et valeur de l'information Géographique. Paris: economica.15p.

**Diwakar P.G., Raghavaswamy V., Saha S.K., Kiran kumar A.S., Dadhwal V.K., Shivakumar S.K. (2013).** Indian remote sensing satellite series and application: A saga of 25years.Bulletin of the national natural resources management system, UNRMS (B)-37. 164p.

**DNSI (1994).** Enquête budget consommation, Ministry of Planning (Bamako). Annuaire statistique de Bamako, Mali. 190p.

**Dubreuil V. (2010).** Geographical information, remote sensing and climatology. In "Geographical Information and climatology" Dir.P Carrega,wiley-ISTE Ltd ; London. pp73-102.

**Francisco E.G., Javier M.R., Ferran M.A. (2013).** "Handbook of spatial remote sensing, Telecan". University of Las Palmas de Gram Canaria, UPC. 8p.

**Fortier J.F. (2007).** *Eichhornia Crassipes.* (Research report) 19p.

**Girard M. C. and Girard C. M. (1999).** Traitement des données de télédétection. Dunod, Paris, 529 p.

**Gopal B. (1987).** Water Hyacinth. Netherlands. Aquatic Plant Studies 60p.

**Hade A. (2002).** Nos lacs- les connaître pour mieux les protéger. Editions Fide, 360p.

**Hassane Younoussou H. (2010).** Prolifération des plantes aquatiques envahissantes sur le fleuve Niger ; état des lieux de la pollution en azote et en phosphore des eaux du fleuve, 22p.

**Holm Lg, Plucknett Dl, Pancho Jv, Herberger J.P. (1977).** The world's worst weeds: distribution and biology. Honolulu: University Press of Hawaii, 609 p.

**IER (2012).** Integrated pest management of aquatic weeds. 6p.

**INSTAT. (2009).** 4ème recensement général de la population et de l'habitat du mali (RGPH-2009): Urbanisation Analyse des résultats définitifs Bamako: Institut Nationale de la Statistique. 57 p.

**Jacquin A. (2010).** Dynamics of savannah vegetation in relation to fire use in Madagascar. Time series analysis of remote sensing images. 146p.

**Garden (2015).** Water hyacinth, *eichhornia crassipes*. https: www.aujardin.info.

**Jean F. (2018).** Contribution à la lutte contre les plantes envahissantes dans le les plans d'eau du Burkina Faso: monitoring du lac de Tengréla par Télédétection. 39p.

**Jimenez and Maricella M. (2016).** Control technique for water hyacinth and other harmful invasive aquatic plants. Cuernaca, Mexico, Instituto de Tecnologia de Agua (IMTA). Accessed october. 22p.

**Justine R. (2017).** Impact of freshwater hyacinth on surface water mobilisation in Burkina Faso, 9 p.

**Khalid A., Abdellatif T., Zakaria A., Mohamed A., Abdelkader El G. (2016).** Contribution of remote sensing and GIS for spatio-temporal monitoring of land use in the West Tarmast catchment. 3[ème] Edition du Colloque International des utilisateurs du SIG, Maroc, Oujda les 22 et 23 Novembre 2016. 31p.

**Komelan Y. (1999).** "L'eutrophisation des retenues d'eau en Côte d'Ivoire et gestion intégrée de leur bassin versant : cas de la Lobo à Daloa" dissertation. 22p.

**Konaté F. (2011).** L'occupation anarchique de l'espace public et les changements de vocation des espaces verts et des places publiques dans le District de Bamako, Revue du Laboratoire de Recherches Biogéographiques et d'Études Environnementales (LaRBE), n°005, ISSN 1812-1403.

**Konecny G. (2003).** Geoinformation: remote sensing, photogrammetry, and geographic information systems. Second edition. Taylor & Francis. 4p.

**Lacombe J.P., Sheeren D. (2007).** Cours de télédétection aérospatiale, télédétection satellitaire, 67p.

**Lillesand T.M., Kifefer RW (1994).** Remote sensing and image interpretation. John wiley and sons, ed.3 ; 750p. New-york.

**Mandi L., Darley J., Barbe J., and Baleux B. (1992).** Essais d'épuration des eaux usées de Marrakech par la jacinthe d'eau (charge organique, Bactérienne et Parasitologique). Revue des sciences de l'eau/ journal of water science n°5, pp.31-33.

**Maurer T. (2013).** How to pan-sharpen images using the gram-schmidt pan-sharpen. International Archives of the Photogrammetry, Remote Sensing and Spatial Information Sciences, XL-1/W1, 6p.

**Maiga F. (2012).** Analyse des externalités négatives de développement urbain du District de Bamako et pollution du fleuve Niger, 46p.

**Millimono S. (2009).** Évaluation des pressions exercées sur les ressources en eau par la ville de Bamako au Mali 8p.

**Nabil B. (2000).** Méthodes de Classification Multicritère : Méthodologie et Applications à l'Aide au Diagnostic Médical. Université libre de Bruxelles institut de statistique et de recherche opérationnelle, 40p.

**Navarro A.L., Phiri G. (2000)** Water Hyacith in Africa and the Middle East. A Survey of problems and solution. Canada. Proceedings of the International Conference on water hyacinth, 16p.

**Ouédraogo R.L. (2000).** Projet pilote de contrôle intégré de *Eichornia Crassipes* ou jacinthe d'eau au Burkina Faso 94p.

**Oumarou, Al Moustapha, Jeanne M-R., S. (2008).** Biogas and compost production from water hyacinth for sustainable development in Sahelian Africa. *VertigO - la revue électronique en sciences de l'environnement* 8 (1).

**Oszwald J., Kouacou Atta J.-M., Kergomard C. and Robin M. (2007).** Representing space to structure time: a remote sensing approach to forest change dynamics in south-eastern Côte d'Ivoire. Revue Télédétection, Vol.7, N°1-2-3-4, pp. 271-282.

**Oloukoi J., Vincent J. Mama, Fulbert B., Agbo J. (2006).** Modélisation de la dynamique de l'occupation des terres dans le département des collines au bénin, Télédétection, vol. 6, no. 4, pp. 305-323.

**Pekkarinen A. (2002).** Image segment-based spectral feature in the estimation of timber volume, remote sensing of environment, n°82 ; pp349-359.

**Pocquet, Inana, Mayako T.O., Antsa R. (2015).** Health and well-being: is hyacinth only a parasitic plant for Côte d'Ivoire? Overblog. 19p.

**Pontius R.G. (1991).** Quantification error versus location in comparison of categorical maps, photogrammetric enginneering and remote sensing, N°66, pp.1011-1016.

**Qaisar M., Zheng P., Siddiqi Mr., Islam E., Azim M., Yousaf H. (2005).** Anatomical studies on water hyacinth (*Eichhornia crassipes* (MART.) Solms) under the influence of textile waste water. J. Zhejlang. Univ.6B (10), p.991-998.

**Rachid B., Yahia El K., Aiman H. (2016).** Apport de la télédétection spatiale à l'étude diachronique de la dynamique de l'occupation du sol dans le bassin versant de l'ouest El Abid (haut Atlas Central, Maroc). 259p.

**Ranarijaona, Her L T., Felice Z., Ainazo H A., George S A. (2013).** Assessment of water hyacinth proliferation in Lake Ravelobe Anharafantsika and restoration plan, vol.13, 1p.

**Rutabagaya J. (2017).** Impact of freshwater hyacinth on surface water mobilisation in Burkina Faso, 25p.

**Sane I. (2001).** Note Synthétique sur la fougère aquatique, Salvania molesta D.S Mitchell et son ennemi naturel spécifique le charançon cyrtobagous salviniae calder and sands, 11 p.

**Saunders P. (2013).** "Purple root water hyacinth is a natural solution to pollution" 41p.

**SIDIBÉ Souleymane, (2017).** Niger River in Mali: Issues and prospects.

**Skupinski G., Don B. T., Weber C. (2009).** Spot multi-date satellite imagery and spatial lametrics in the study of urban and suburban change. Le cas de la basse vallée de Bruche (Bas-

Rhin) European Journal of Geography. URL: http://cybergeo.revues.org/21995; DOI: 10.4000/cybergeo.21995. France.

**Shalaby A., Tateishi R. (2007).** Remote sensing and GIS for mapping and monitoring land cover and land-use changes in the Northwestern coastal zone of Egypt. Applied Geography, N°27, pp.28-41.

**Toko I. (2014).** Factors determining the fragmentation of forest ecosystems: the case of the islands of dense dry forest in the Monts kouffé classified forest and its periphery in Benin (Phd), Abomey-Calavi, Benin, 231p.

**Toukiloglou P. (2007).** Comparison of AVHRR, MODIS and vegetation for land cover mapping and drought monitoring at 1km spatial resolution. PhD thesis, Cranfield University, England. 515p.

**Traore H. (2006).** Un problème environnemental transformé en source de revenu, cas de la jacinthe d'eau douce en compost chez des jardiniers de Bamako, 7p.

**Yacine K. (2018).** Geographic information system course. Université Larbi Ben M'Hidi-OEB-Algérie. 4p.

**Yuhendra A.B., Alimuddina I.C., Sumantyoa J.T.S., Hiroaki K. (2012).** Assessment of pan-sharpening methods applied to image fusion of remotely sensed multi-band data. International Journal of Applied Earth Observation and Geoinformation, 18, pp165-175.

## WEBOGRAPHY

**https://www.afdb.org Accessed on 02 April 2018**
.www.fao.org/3/a-y4270f.pdf. Accessed on 12 May 2018

www.abfn-mali.org/.../Rapport%20Identification%20%20des%20zones%20infestées%...
Accessed on 12 May 2018

www.abfn-mal20. Consultétude%20sur%20la%20lutte%20contre%20la%20jacinthe%20.
Accessed on 18 July 2018

**https://journals.openedition.org/cybergeo/22707 consulted on 13/10/2018** at 00h34mm22s

https://maliactu.net/mali-fleuve-niger-au-mali-enjeux-et-perspectives/ 29/10/2018.

APPENDICES

**Appendix 1: Average daily evolution curves for the River Niger in the Bamako district in 2000.**

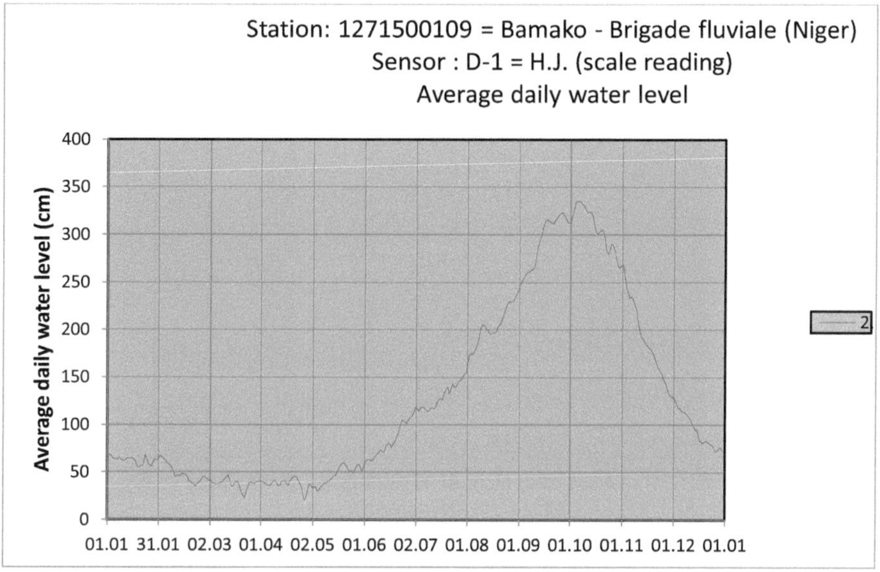


**Source: DNH (Direction Nationale d'Hydraulique du Mali).**

**Appendix 2: Average daily evolution curves for the River Niger in the Bamako district in 2006.**

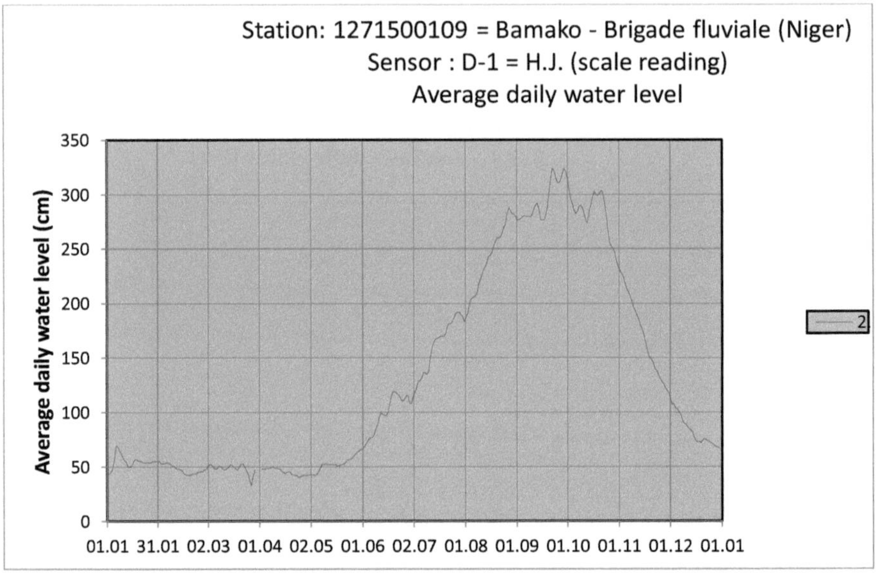


**Source: DNH (Direction Nationale d'Hydraulique du Mali).**

**Appendix 3: Average daily evolution curves for the River Niger in the Bamako district in 2012.**

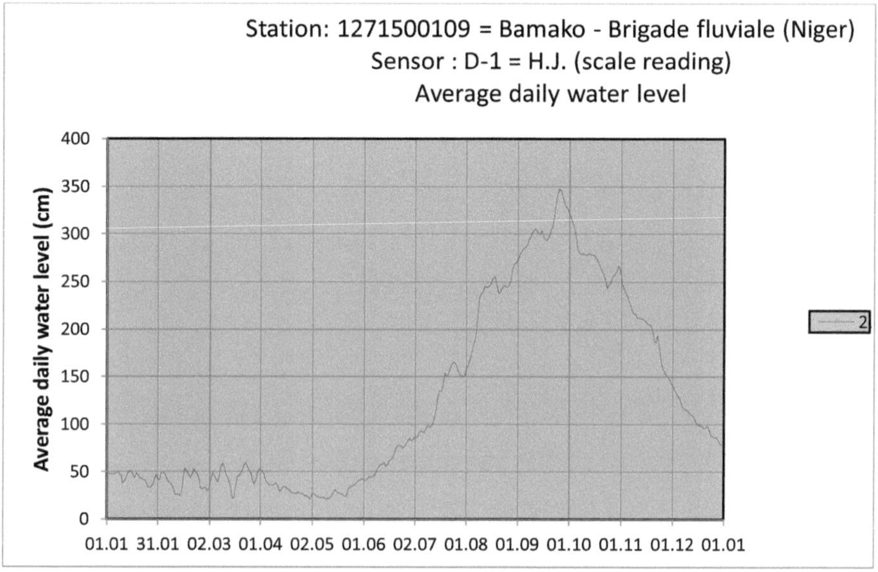


**Source: DNH (Direction Nationale d'Hydraulique du Mali).**

**Appendix 4: Average daily evolution curves for the River Niger in the Bamako district in 2018.**

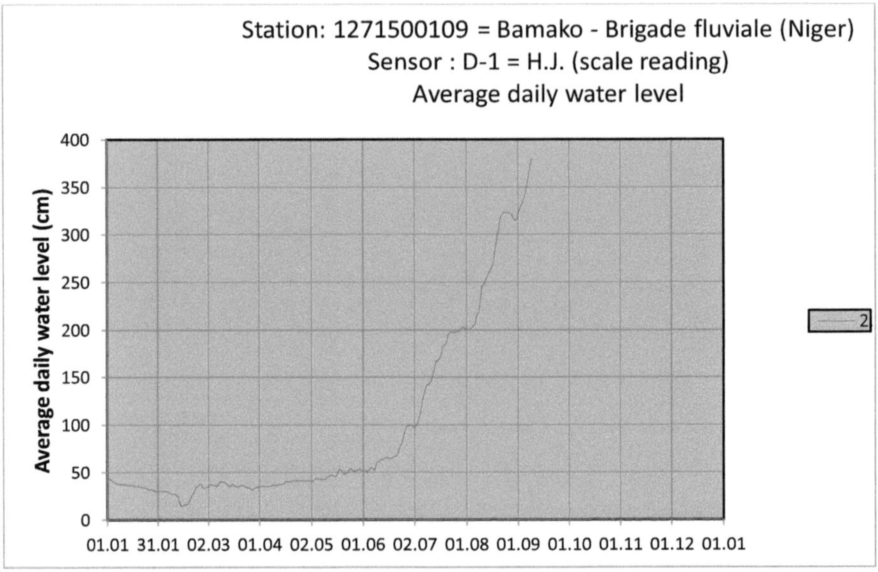


**Source: DNH (Direction Nationale d'Hydraulique du Mali).**

**Appendix 5: Sample of coordinates surveyed in the field**

| N° | Contact details | |
|---|---|---|
| | Longitude (X) | Latitude (Y) |
| 1 | -7,99135 | 12,6333 |
| 2 | -7,99144 | 12,63337 |
| 3 | -7,9915 | 12,63338 |
| 4 | -7,9917 | 12,63333 |
| 5 | -7,99179 | 12,6333 |
| 6 | -7,99144 | 12,63325 |
| 7 | -8,00124 | 12,62901 |
| 8 | -7,99378 | 12,63298 |
| 9 | -7,99375 | 12,63286 |
| 10 | -7,99386 | 12,63273 |
| 11 | -7,99395 | 12,63258 |
| 12 | -7,99408 | 12,63263 |
| 13 | -7,99421 | 12,63265 |
| 14 | -7,99442 | 12,6323 |
| 15 | -7,99658 | 12,63139 |
| 16 | -7,9966 | 12,63132 |
| 17 | -7,99672 | 12,63124 |
| 18 | -7,99685 | 12,63123 |
| 19 | -7,97186 | 12,61803 |
| 20 | -7,97732 | 12,61909 |
| 21 | -7,99886 | 12,62936 |
| 22 | -7,99889 | 12,62936 |
| 23 | -8,00277 | 12,62796 |
| 24 | -8,00275 | 12,62791 |
| 25 | -8,00278 | 12,62785 |
| 26 | -8,00283 | 12,6279 |
| 27 | -8,00286 | 12,62792 |
| 28 | -8,00289 | 12,62794 |
| 29 | -7,96518 | 12,63733 |
| 30 | -7,96467 | 12,62255 |
| 31 | -7,96669 | 12,62113 |
| 32 | -7,96626 | 12,6375 |
| 33 | -7,96776 | 12,62039 |
| 34 | -7,98849 | 12,62648 |
| 35 | -7,9885 | 12,62652 |
| 36 | -7,99144 | 12,63325 |
| 37 | -7,98823 | 12,6266 |
| 38 | -7,98815 | 12,62663 |

MIX
Papier aus verantwortungsvollen Quellen
Paper from responsible sources
FSC® C105338

Printed by Books on Demand GmbH, Norderstedt / Germany